Memoirs of the American Mathematical Society

Number 429

Etsuko Bannai

Positive definite unimodular lattices with trivial automorphism groups

Published by the
AMERICAN MATHEMATICAL SOCIETY
Providence, Rhode Island, USA

May 1990 · Volume 85 · Number 429 (second of 3 numbers)

1980 *Mathematics Subject Classification* (1985 *Revision*).
Primary 10C.

Library of Congress Cataloging-in-Publication Data

Bannai, Etsuko, 1944–
    Positive definite unimodular lattices with trivial automorphism groups/Etsuko Bannai.
    p. cm. – (Memoirs of the American Mathematical Society, ISSN 0065-9266; no. 429)
    "May 1990."
    Includes bibliographical references.
    "Volume 85 number 429."
    ISBN 0-8218-2491-0
    1. Modular lattices. 2. Automorphisms. I. Title. II. Series.
QA3.A57 no. 429
[QA171.5]
510 s–dc20   90-31824
[511.3$'$3]   CIP

Subscriptions and orders for publications of the American Mathematical Society should be addressed to American Mathematical Society, Box 1571, Annex Station, Providence, RI 02901-1571. *All orders must be accompanied by payment.* Other correspondence should be addressed to Box 6248, Providence, RI 02940-6248.

SUBSCRIPTION INFORMATION. The 1990 subscription begins with Number 419 and consists of six mailings, each containing one or more numbers. Subscription prices for 1990 are $252 list, $202 institutional member. A late charge of 10% of the subscription price will be imposed on orders received from nonmembers after January 1 of the subscription year. Subscribers outside the United States and India must pay a postage surcharge of $25; subscribers in India must pay a postage surcharge of $43. Each number may be ordered separately; *please specify number* when ordering an individual number. For prices and titles of recently released numbers, see the New Publications sections of the NOTICES of the American Mathematical Society.

BACK NUMBER INFORMATION. For back issues see the AMS Catalogue of Publications.

MEMOIRS of the American Mathematical Society (ISSN 0065-9266) is published bimonthly (each volume consisting usually of more than one number) by the American Mathematical Society at 201 Charles Street, Providence, Rhode Island 02904-2213. Second Class postage paid at Providence, Rhode Island 02940-6248. Postmaster: Send address changes to Memoirs of the American Mathematical Society, American Mathematical Society, Box 6248, Providence, RI 02940-6248.

TABLE OF CONTENTS

INTRODUCTION 1

CHAPTER   PAGE

   I.   PRELIMINARIES. 7

       §1. Quadratic lattices and their mass formulas. 7

       §2. Hermitian lattices and their mass formulas. 8

  II.   GENERAL THEORY 11

       §3. Lattices with non-trivial automorphisms whose minimal
           polynomials are reducible. 11

       §4. Lattices with non-trivial automorphisms whose minimal
           polynomials are irreducible. 18

 III.   LOCAL DENSITIES 22

       §5. Local densities of hermitian lattices. 22

       §6. List of quadratic lattices over 2-adic integers and their
           local densities. 34

 IV.   ESTIMATIONS 38

       §7. The estimation of $\omega_{R(2)}/\omega(L)$ 41

       §8. The estimation of $\omega_{R(q)}/\omega(L), q \neq 2$ 55

       §9. The estimation of $\omega_{IR(q)}/\omega(L)$ 64

   V.   PROOF OF THE THEOREMS 67

       §10. Proof of the theorems. 67

LIST OF REFERENCES 69

Abstract

In this paper the following two theorems are proved.

Theorem 1. If m is sufficiently large, then there exists a lattice with trivial automorphism group (i.e. $\{\pm 1\}$) in every genus of positive definite unimodular $\mathbb{Z}$-lattices of rank m. More precisely, we have

(i) the above assertion holds if $m \geq 43$ for odd unimodular lattices,

(ii) the above assertion holds if $m \geq 144$ for even unimodular lattices.

Theorem 2. Let ω be the mass of the given genus of positive definite unimodular $\mathbb{Z}$-lattices of rank m and let ω' be the mass of all the classes with nontrivial automorphisms. Then the ratio of the masses ω'/ω is bounded above by $30(\sqrt{2\pi})^m/\Gamma(\frac{m}{2})$ for odd unimodular lattices of rank $m \geq 43$ and by $2^{m+1}(\sqrt{2\pi})^m/\Gamma(\frac{m}{2})$ for even unimodular lattices of rank $m \geq 144$. In particular, this ratio ω'/ω approaches 0 very rapidly as m increases.

The existence of unimodular $\mathbb{Z}$-lattices with trivial automorphism group was not previously known. Theorems 1 and 2 above show that there are many such lattices in a given genus of positive definite unimodular $\mathbb{Z}$-lattices of sufficiently large rank.

Theorem 1 is a consequence of Theorem 2. As for the proof of Theorem 2, for every lattice in the given genus with nontrivial automorphisms we can construct a positive definite quadratic lattice and a totally positve definite hermitian lattice over $\mathbb{Q}(\varsigma)$, where ς is a primitive q-th root of unity, where $q = $ either 4 or a prime number. Those lattices satisfy some good conditions. By means of Siegel's mass formulas (both orthogonal and hermitian cases) for these lattices, we evaluate ω' explicitly and obtain the above upper bounds for ω'/ω.

Key words and phrases: quadratic form, unimodular lattice, hermitian form, mass formula, local density.

INTRODUCTION

The main objective of this thesis is to prove the following two theorems.

Theorem 1. If m is sufficiently large, then there exists a lattice with trivial automorphism group (i.e. $\{\pm 1\}$) in every genus of positive definite unimodular integral lattices of dimension m. More precisely, (i) the assertion holds if $m \geq 43$ for odd unimodular lattices, (ii) the assertion holds if $m \geq 144$ for even unimodular lattices.

Theorem 1 is obtained from the following stronger theorem.

Theorem 2. Let ω be the mass of the given genus of positive definite unimodular lattices of rank m and let ω' be the mass of all the classes in the given genus with nontrivial automorphisms. Then the ratio of the mass ω'/ω is bounded above by $30(\sqrt{2\pi})^m/\Gamma(\frac{m}{2})$ for odd unimodular lattices of dimension $m \geq 43$ and by $2^{m+1}(\sqrt{2\pi})^m/\Gamma(\frac{m}{2})$ for even unimodular lattices of dimension $m \geq 144$. In particular, this ratio ω'/ω approaches 0 very rapidly as m increases.

Here is a brief historical background concerning this problem. The existence of lattices with trivial automorphism group is known. O'Meara [9, 1975] gave an algorithm to construct such a lattice starting from any given lattice. In this process the discriminants of the lattices increase in each step. Biermann [1, 1981] proved the existence of a lattice with trivial automorphism group in every genus of positive definite integral lattices of any dimension with sufficiently large discriminant. In his proof the fact that the discriminant is very large is crucial. We are, instead, interested in lattices with small discriminant. It seems that the existence of any unimodular lattice with trivial automorphism group has not been known. This was, however, anticipated in [6]. For a treatment over localizations of polynomial rings, see [13].

On the other hand, Watson [17, 1979] has shown the existence of an indecomposable lattice in every genus of positive definite integral lattices if the dimension $m \geq 14$. Clearly a lattice having trivial automorphism group is indecomposable, but not vice-versa. But his idea of estimating the mass of decomposable lattices in the given genus is, however, very

---

Received by the editors January 31, 1989.

useful for our study. (Contrarily to Watson, Biermann estimated the number of the classes in the genus with non-trivial automorphism groups).

Our Theorems 1 and 2 show that there are an abundance of lattices with trivial automorphism group in every genus of positive definite unimodular integral lattices if the dimension is not too small, and furthermore the ratio

$$\frac{\text{mass of classes in } G_L \text{ with trivial automorphism group}}{\text{mass of } G_L}$$

approaches to 1 as the dimension increases.

It is important to note, however, that the *explicit construction* of a lattice with trivial automorphism group is still an open problem!

Now we will give an outline of the proof of Theorems 1 and 2.

Let L be a positive definite unimodular integral lattice of rank m with a bilinear form $B(-,-)$. Let G_L be the genus of L. Define $\omega(L) = \sum\limits_{cls\, M \subseteq G_L} \frac{1}{|O(M)|}$ and $\omega'(L) = \sum\limits_{\substack{cls\, M \subseteq G_L \\ |O(M)| > 2}} \frac{1}{|O(M)|}$. The basic strategy of the proof is to estimate this ratio $\omega'(L)/\omega(L)$.

Let M be a lattice in G_L with non-trivial automorphisms. If the order of the automorphism group of M is divisible by some odd prime number q, then there exists an isometry g of order q. Since $g^q = 1$, the minimal polynomial of g must divide $x^q - 1$. Therefore the minimal polynomial of g is $x^q - 1$ (reducible) or $x^{q-1} + \cdots x + 1$ (irreducible). If the minimal polynomial is $x^q - 1$, then $q \leq m$ and if the minimal polynomial is $x^{q-1} + \cdots + x + 1$ then $q - 1 \leq m$. Next, consider the case when the order of the automorphism group is a power of 2. If M has an isometry g of order 2 which is not trivial then the minimal polynomial is $x^2 - 1$. If every isometry of M of order 2 is trivial and $4 \mid |O(M)|$, then any isometry of order 4 has minimal polynomial $x^2 + 1$.

Thus we have seen that the lattices with nontrivial automorphisms are either (or both) of the following two types.

Type $R(q)$. Lattice with an isometry of order a prime number $q \leq m$ (q is odd or 2) whose minimal polynomial is reducible, i.e., $x^q - 1$.

Type $IR(q)$. Lattice with an isometry of order $q \leq m + 1$ (where q is an odd prime number or 4) whose minimal polynomial is irreducible i.e., if q is an odd prime then $x^{q-1} + \cdots + x + 1$ and if $q = 4$ then $x^2 + 1$.

Let

$$\omega_{R(q)} = \sum_{\substack{cls\, M \subseteq G_L \\ type\, R(q)}} \frac{1}{|O(M)|} \quad \text{and} \quad \omega_{IR(q)} = \sum_{\substack{cls\, M \subseteq G_L \\ type\, IR(q)}} \frac{1}{|O(M)|}.$$

Then we have

$$\omega'(L) \leq \sum_{\substack{q \, prime \\ 2 \leq q \leq m}} \omega_{R(q)} + \sum_{\substack{q \, odd \, prime \\ (q-1)|m \, or \\ q=4}} \omega_{IR(q)}$$

Remark: If G_L contains a lattice of type $IR(q)$ then the rank m has to be an even number. Therefore if m is an odd number then we only need to consider the type $R(q)$.

We are going to estimate $\omega_{R(q)}/\omega(L)$ and $\omega_{IR(q)}/\omega(L)$ separately. The idea of dividing lattices in the given genus into those two types is due to Biermann [1]. He introduced hermitian structures to the lattices of type $IR(q)$. In this paper we go further and also introduce hermitian structures to the lattices of type $R(q)$.

Type $R(q)$. Let M be a lattice of type $R(q)$ in G_L. Let $G = \langle g \rangle$ be the cyclic group of order q generated by $g \in O(M), \mathbb{Q}G$ be the group algebra and $\Lambda = \mathbb{Z}G$ be the group ring. Let ς be a primitive q-th root of unity. Then $\mathbb{Q}G$ is isomorphic to $\mathbb{Q} \times \mathbb{Q}(\varsigma)$ under the mapping which corresponds g to $(1, \varsigma)$. Let $\Gamma = \mathbb{Z} \times \mathbb{Z}[\varsigma] \subset \mathbb{Q} \times \mathbb{Q}(\varsigma)$. Then $\Lambda = \mathbb{Z}G$ is injected into Γ by the above isomorphism. We often identify $\mathbb{Q}G$ and $\mathbb{Q} \times \mathbb{Q}(\varsigma)$. (Note that if $q = 2$ then $\varsigma = -1$ and $\mathbb{Q}(\varsigma) = \mathbb{Q}$.)

The following arguments are shown in Quebbemann [11,§2] in a more general way, as we need only the rational number and rational integer cases, so we provide the proof below.

Let $W = \mathbb{Q}M$, $W_0 = (\sum_{i=0}^{q-1} g^i)W$ and $W_1 = (g-1)W$. Then $W = W_0 \perp W_1$ (with respect to the bilinear form B). The group algebra $\mathbb{Q}G$ acts on W, and M is a Λ-module. Let $h(x, y) = \sum_{i=0}^{q-1} B(g^{-i}x, y)g^i$ for $x, y \in W$. Then h is an hermitian form with respect to the involution sending g to g^{-1} (or equivalently $(1, \varsigma)$ to $(1, \overline{\varsigma})$ where $\overline{\varsigma}$ is the complex conjugate of ς). Clearly $h(x, y) \in \Lambda$ for any $x, y \in M$ and M is an hermitian Λ-lattice. Let us denote the hermitian structure of W and M by $\tilde{W}$ and $\tilde{M}$, respectively. Let $\Gamma \tilde{M} = M_0 \times M_1$ where $M_0 = (\mathbb{Z} \times \{0\})\tilde{M}$ and $M_1 = (\{0\} \times \mathbb{Z}[\varsigma])\tilde{M}$. Let B_0 be the restriction of h to M_0 and h_1 the restriction of h to M_1. We will see in §3 that (M_0, B_0) is a positive definite integral $\mathbb{Z}$-lattice with the bilinear form B_0 and (M_1, h_1) is a totally positive definite hermitian $\mathbb{Z}[\varsigma]$-lattice with the hermitian form h_1. Let us denote the structure (M_1, h_1) by $(\mathcal{M}_1, h_1)$. Thus, while the original $\mathbb{Z}$-lattice (M, B) may be orthogonally indecomposable, the Γ-lattice $(\Gamma \tilde{M}, h)$, which is slightly bigger, has an important orthogonal (with h) decomposition. This is the central reason for dealing with Γ. We are going to estimate $\omega_{R(q)}$ through Siegel's mass formula for integral $\mathbb{Z}$-lattices M_0 and the hermitian $\mathbb{Z}[\varsigma]$-lattices $\mathcal{M}_1$. (Note that if $q = 2$, then $\mathcal{M}_1$ is also a positive definite integral $\mathbb{Z}$-lattice with bilinear form h_1. However, in this special case we shall denote $\mathcal{M}_1$ by M_1 and h_1 by B_1.)

Let λ be a prime element in $\mathbb{Z}[\varsigma]$ above q. Let $I = q\Lambda + (1-g)\Lambda \subset \mathbb{Z}G$, then I corresponds to $q\mathbb{Z} \times \lambda\mathbb{Z}[\varsigma]$ in $\Gamma = \mathbb{Z} \times \mathbb{Z}[\varsigma]$.

Let $V_0(M_0) = M_0/qM_0$, $V_1(M_1) = M_1/\lambda M_1$ be the vector spaces over the finite field $\mathbb{F}_q$ of q elements. Let b_0 and b_1 be the bilinear form induced by $B_0(\bmod\ q\mathbb{Z})$ and h_1 $(\bmod\ \lambda\ \mathbb{Z}[\varsigma])$, respectivley. Let $V_0'(M_0) = V_0(M_0)/V_0(M_0)^\perp$ and $V_1'(M) = V_1(M_1)/V_1(M_1)^\perp$ be the vector spaces over the field $\mathbb{F}_q$ with the nonsingular bilinearform b_0' and b_1' induced by b_0 and b_1, respectively. (Here a bilinear form b on a vector space V is said to be non-singular if $b(x,y) = 0$ for any $x \in V$ implies $y = 0$.) Then it is shown in Quebbemann [11], Satz 2.1, that $(V_0'(M_0), b_0') \cong (V_1'(M_1), b_1')$. See also Proposition 3 in §3 below.

Let $q \leq m$ be a prime number, and let $r \geq 1$ and $\rho \geq 0$ be integers satisfying $r(q-1) \leq m-1$ and $\rho \leq min(r, m_0)$, where $m_0 = m - r(q-1)$ (if $q = 2$ assume $r \leq m_0$, i.e. , $r \leq [\frac{m}{2}]$). For each such q, r and ρ, define $L(q, r, \rho)$ to be the set of all the lattices M in G_L such that M is of type $R(q)$, $rank_{\mathbb{Z}[\varsigma]}M_1 = r$ (therefore $rank_{\mathbb{Z}}M_0 = m_0$), and $dim_{\mathbb{F}_q}V_0'(M_0) = dim_{\mathbb{F}_q}V_1'(M_1) = \rho$.

Define $G(q, r, \rho)$ to be the set of all the pairs of genera $(G_{N_0}, G_{\mathcal{N}_1})$ satisfying the following conditions:

If $q = 2$, then

(i) N_0 and $\mathcal{N}_1 = N_1$ are positive definite integral $\mathbb{Z}$-lattices of rank m_0 and r, respectively.

(ii) $dN_0 = 2^{m_0-\rho}$, $dN_1 = 2^{r-\rho}$.

(iii) Let y be in the vector space $\mathbb{Q}N_i$, $i = 0, 1$, then $B_i(x, y) \in 2\mathbb{Z}$ for all $x \in N_i$ implies $y \in N_i$, where B_i is the bilinear form of N_i.

(iv) $(V_0'(N_0), b_0')$ is isometric to $(V_1'(N_1), b_1')$.

(v) $dim_{\mathbb{F}_q}V_i'(N_i) = \rho$, $i = 0, 1$.

If $q \neq 2$, then

(i) N_0 is a positive definite integral $\mathbb{Z}$-lattice of rank m_0, and $\mathcal{N}_1$ is a totally positive definite hermitian $\mathbb{Z}[\varsigma]$-lattice of rank r.

(ii) $dN_0 = q^{m_0-\rho}$, $N_{K/\mathbb{Q}}(\delta \mathcal{N}_1) = q^{(r-\rho)/2}$

(iii) Let $y \in \mathbb{Q}N_0$, then $B_0(x, y) \in q\mathbb{Z}$ for all $x \in N_0$ implies $y \in N_0$. Let $y \in \mathbb{Q}(\varsigma)\mathcal{N}_1$, then $h_1(x, y) \in \lambda\mathbb{Z}[\varsigma]$ for all $x \in \mathcal{N}_1$ implies $y \in \mathcal{N}_1$. Here B_0 is the bilinear form of N_0 and h_1 is the hermitian form of $\mathcal{N}_1$.

(iv) $(V_0'(N_0), b_0')$ is isometric to $(V_1'(\mathcal{N}_1), b_1')$.

(v) $dim_{\mathbb{F}_q}V_0'(N_0) = dim_{\mathbb{F}_q}V_1'(\mathcal{N}_1) = \rho$.

(vi) If L is even then $\mathcal{N}(N_0) \subseteq 2\mathbb{Z}$.

Then we show in §3 that for any lattices M_0 and $\mathcal{M}_1$ constructed from a lattice $M \in L(q, r, \rho)$, the pair $(G_{M_0}, G_{\mathcal{M}_1})$ is contained in $G(q, r, \rho)$ and

$$\sum_{\substack{cls\ K \subseteq L(q,r,\rho) \\ \Gamma\tilde{K} \cong \Gamma\tilde{M}}} \frac{1}{|O(K)|} \leq \frac{|I(V_0'(M_0), V_1'(\mathcal{M}_1))|}{|Aut\ M_0||Aut\ \mathcal{M}_1|}.$$

Therefore

$$\sum_{cls\ K \subseteq L(q,r,\rho)} \frac{1}{|O(K)|} \leq \sum_{(G_{N_0}, G_{\mathcal{N}_1}) \in G(q,r,\rho)} |I(V_0'(N_0), V_1'(\mathcal{N}_1))|\omega(N_0)\omega(\mathcal{N}_1)$$

where $Aut\ M_0$ is the orthogonal group of M_0 with respect to B_0, $Aut\ \mathcal{M}_1$ is the unitary group of $\mathcal{M}_1$ with respect to h_1 (if $q = 2$ then $\mathcal{M}_1 = M_1$, $h_1 = B_1$, and $Aut\ \mathcal{M}_1$ is the orthogonal group of M_1 with respect to B_1), $\omega(N_0)$ is the mass of N_0, $\omega(\mathcal{N}_1)$ is the mass of $\mathcal{N}_1$, and $I(V_0'(N_0), V_1'(\mathcal{N}_1))$ is the set of all the isometries from $(V_0'(N_0), b_0')$ to $(V_1'(\mathcal{N}_1), b_1')$ which depends only on the pair $(G_{N_0}, G_{\mathcal{N}_1})$ (if $q = 2$ then $\mathcal{N}_1 = N_1$ is an integral $\mathbb{Z}$-lattice). The orders of the orthogonal groups over finite fields are explicitly known (see e.g. [4]). Finally we will show that

$$\omega_{R(q)} < \sum_{r=1}^{[\frac{m-1}{q-1}]} \sum_{\rho=0}^{min(r,m_0)} \sum_{(G_{N_0}, G_{\mathcal{N}_1}) \in G(q,r,\rho)} |I(V_0'(N_0), V_1'(\mathcal{N}_1)|\omega(N_0)\omega(\mathcal{N}_1).$$

Using the classification of quadratic lattices over q-adic integers by O'Meara and the classification of hermitian lattices over local fields by Jacobowitz and Shimura we show in Chapter III that the number of the pairs of genera in $G(q, r, \rho)$ is bounded by a constant (independent of m, q, r and ρ).

Type $IR(q)$. Let M be a lattice in G_L of type $IR(q)$ with the isometry g. The following argument for the lattices of Type $IR(q)$ is based on Biermann [1]. He did not define the hermitian form explicitly. Here we will give a explicit definition.

Let M be a lattice in G_L of type $IR(q)$, g the isometry of M. Let $W = \mathbb{Q}M$ and define the action of $\mathbb{Q}(\varsigma)$ on W by $\varsigma \cdot x = g(x)$ for all $x \in W$. Then W is a $\mathbb{Q}(\varsigma)$ vector space and M is a $\mathbb{Z}[\varsigma]$-module. For $q \neq 4$ define $h(x, y) = \sum_{i=0}^{q-1} B(g^{-i}x, y)\varsigma^i$ and for $q = 4, h(x, y) = \frac{1}{2} \sum_{i=0}^{3} B((q^{-i}x, y)\varsigma^i$. Then h is a totally positive definite hermitian form. Clearly $h(x, y) \in \mathbb{Z}[\varsigma]$ for $x, y \in M$. We show in §4 that with this hermitian form M has a $\lambda\mathbb{Z}[\varsigma]$- modular hermitian structure if $q \neq 4$, and a unimodular hermitian lattice structure if $q = 4$.

Let us denote the hermitian $\mathbb{Z}[\varsigma]$-lattice structure of M by $\mathcal{M}$. Then we will see in §4 the following inequalities:

$$\omega_{IR(q)} \leq \omega(\mathcal{M}) \qquad \text{for } q \neq 4$$
$$\omega_{IR(4)} \leq \sum_{G_{\mathcal{N}}} \omega(\mathcal{N}) \text{for } q = 4$$

where the summation in the second inequality is over the set of genera of totally positive definite unimodular hermitian $\mathbb{Z}[\sqrt{-1}]$-lattices $\mathcal{N}$ of rank $r = \frac{m}{2}$.

The classification of hermitian lattices over local fields by Jacobowitz and Shimura shows that there is exactly one genus of totally positive definite $\lambda\mathbb{Z}[\varsigma]$-modular hermitian $\mathbb{Z}[\varsigma]$ lattices of rank $r = \frac{m}{q-1}$ with $q \neq 4$, and exactly two genera of totally positive definite unimodular hermitian $\mathbb{Z}[\sqrt{-1}]$-lattices of rank $r = \frac{m}{2}$ of norm $\mathbb{Z}[\sqrt{-1}]$ and exactly one genus of such lattices of norm $2\mathbb{Z}[\sqrt{-1}]$.

In Chapter IV we give the actual evaluations of $\omega_{R(q)}/\omega(L)$ and $\omega_{IR(q)}/\omega(L)$. In Chapter V we complete the proof of the theorems.

In Chapter III we give some formulas of local densities of quadratic or hermitian lattices. Formulas for the quadratic cases have already been studied extensively by Siegel [15], Watson [16] and others (e.g. Pall [10]). We present part of their results here for the reader's convenience. For the hermitian case, it would seem, as of now, that formulas have been calculated only for the unimodular lattices at unramified primes (see Braun [3], Rehmann [12]). Here we obtain formulas for certain lattices at ramified primes.

In Theorem 2, the conditions on rank ($m \geq 43$ for odd unimodular case and $m \geq 144$ for even unimodular case) seem to be the best possible using our method. The ratio $\omega_{R(2)}/\omega(L)$ has the same order as that of the function $(\sqrt{2\pi})^m/\Gamma(\frac{m}{2})$ in the odd case and $2^m(\sqrt{2\pi})^m/\Gamma(\frac{m}{2})$ in the even case. By calculation we prove that the ratio $\omega_{R(2)}/\omega(L)$ is the dominant term in $\omega'(L)/\omega(L)$. If $m \leq 42$ (resp. $m \leq 136$) then our method can not give a good estimation.

Also using the proof of Theorem 2 we can show the following Theorem 3.

Theorem 3. If $m \geq 43$, then there exists a lattice whose full automorphism group has order a power of 2 in every genus of positive definite unimodular lattices of rank m.

CHAPTER I
PRELIMINARIES

In this chapter we will give the definitions and the basic concepts we need in the proof of the theorems.

Let W be a vector space over a number field E. Let S be the ring of algebraic integers of E. We call a S-submodule M of W a *lattice* in W if there is a base $x_1, \cdots, x_m$ for W such that $M \subseteq Sx_1 + \cdots + Sx_m$. We say that M is a lattice on W if M generates W over E, and in such a case m is the *rank* rank $_S(M)$ of M. Unexplained notations and terminology may be found in O'Meara 's book [8].

§1. Quadratic lattices and their mass formula.

In this section our field E is the rational number field $\mathbb{Q}$.

Let W be a quadratric space of dimension m. Let Q be the quadratic form on W and B be the associated bilinear form on $W \times W$, i.e., $Q(x+y) = Q(x)+Q(y)+2B(x,y)$. Let L be a lattice on W. Then there is a base $x_1, \cdots x_m$ of W such that $L = \mathbb{Z}x_1, + \cdots + \mathbb{Z}x_m$. Define the *discriminant* dL of L to be the determinant of the matrix $(B(x_i, x_j))$. Then dL is independent of the choice of the base. *Scale* sL of L is the $\mathbb{Z}$-module generated by the subset $B(L, L)$ of $\mathbb{Q}$. Define the *norm* nL to be the $\mathbb{Z}$-module generated by $Q(L) \subseteq B(L, L)$. We say L is *unimodular* if $dL = \pm 1$ and $sL = \mathbb{Z}$. We define a unimodular lattice to be *odd* if $nL = \mathbb{Z}$ and *even* if $nL = 2\mathbb{Z}$.

Let p be a prime spot (prime number in $\mathbb{Z}$ or infinite prime). Let $\mathbb{Q}_p$ be the completion of the field $\mathbb{Q}$ at p. For a finite prime $p, \mathbb{Z}_p$ denotes the ring of p-adic integers. Let $W_p = W \otimes_{\mathbb{Q}} \mathbb{Q}_p$ and $L_p = L \otimes_{\mathbb{Z}} \mathbb{Z}_p$. Then the bilinear form B induces a bilinear form on $W_p \times W_p$. We define the *genus* G_L of the lattice L to be the set of all lattices M on W with the following property: for each finite prime p there exists an isometry σ_p in $O(W_p)$ such that $M_p = \sigma_p L_p$. We say a lattice M on W is in the *class* of $L, clsL$, if there exists an isometry σ in $O(W)$ such that $M = \sigma L$. It is known that the number of isometric classes in a genus is finite (see O'Meara [8],§103).

If B is a positive definite bilinear form, then the orthogonal group $O(M)$ of any lattice M on W is a finite group.

7

Now we assume that the quadratic space W has positive definite bilinear form B and the lattice L on W is *integral*, i.e., $B(L, L) \subset \mathbb{Z}$. Define the *mass* of the lattice L by

$$w(L) = \sum_{cls\, M \subseteq G_L} \frac{1}{|O(M)|}.$$

Let μ be a positive integer. Let $A_{p^\mu}(L, L)$ be the number of $\mathbb{Z}$ linear maps $\sigma : L \longrightarrow L$ which are distinct modulo $p^\mu L$ and satisfy $B(\sigma x, \sigma y) \equiv B(x, y)\, mod\, p^\mu$ for every x and y in L. Then it is known that the limit

$$\lim_{\mu \to \infty} p^{-\frac{1}{2}\mu m(m-1)} A_{p^\mu}(L, L)$$

exists and stabilizes for μ sufficiently large (see Siegel [15], Hilfssatz 13). We define *local density* of L at p by

$$\alpha_p(L) = \lim_{\mu \to \infty} \frac{1}{2} p^{-\frac{1}{2}\mu m(m-1)} A_{p^\mu}(L, L).$$

Then we have the following mass formula due to Siegel (see [15], §10).

(1)
$$w(L) = \frac{2 \prod\limits_{i=1}^{m} \Gamma(\frac{i}{2})(dL)^{\frac{m+1}{2}}}{\pi^{\frac{m(m+1)}{4}} \prod\limits_{p\, \text{finite}} \alpha_p(L)}.$$

where $\Gamma(x)$ denotes the gamma function and $m > 1$.

§2. Hermitian lattices and their mass formula.

In this section our field E is $\mathbb{Q}(\varsigma)$ where ς is a primitive q-th root of unity with $q = 4$ or an odd prime number. Let K be the maximal real sub-field in E (i.e., $K = E \cap \mathbb{R}$). Let W be a hermitian space over E of dimension r with respect to the complex conjugation of E. Let h be the hermitian form on W. We assume $h : W \times W \longrightarrow E$ is linear in the first component, $h(x, y) = \overline{h(y, x)}$, and $h(x, \alpha y) = \overline{\alpha}\, h(x, y)$ for any $\alpha \in E$, where $\overline{\alpha}$ is the complex conjugate of α. Let M be a lattice on W. Then there exists a base $x_1, \cdots x_r$ of W and fractional ideals $A_1, \cdots A_r$ of E such that

$$M = A_1 x_1 + \cdots + A_r x_r.$$

Let $d_h(x_1, \cdots, x_m)$ be the determinant of the matrix $(h(x_i, x_j))$. Define the *discriminant ideal* δM by

$$\delta M = (A_1 \overline{A}_1) \cdots (A_r \overline{A}_r) d_h(x_1, \cdots, x_r).$$

The *scale* sM of M is the S-module generated by $h(M, M)$. Define the *norm* nM of M to be the S-module generated by the subset $\{h(x, x) \mid x \in M\}$ of $E = \mathbb{Q}(\varsigma)$. We say

M is A—*modular* if $sM = A$ and $\delta M = A^r$. Particularly, M is *unimodular* if $sM = S$ and $\delta M = S$.

Let R be the ring of algebraic integers and P a prime ideal in K or an infinite prime spot of K. Let K_P be a completion of K at P and R_P be the ring of integers of K_P. We define the following:

$$E_P = E \otimes_K K_P, \; S_P = S \otimes_R R_P,$$

$$W_P = W \otimes_E E_P \text{ and } M_P = M \otimes_S S_P$$

Then complex conjugation induces an involution of E_P which leaves the elements of K_P invariant. There is a unique extension of h on $W_P \times W_P \longrightarrow E_P$ at each P. We defined the *genus* G_M of the lattice M to be the set of all lattices N on W with the following property: for each finite prime P there exists an isometry σ_P in the local unitary group $U(W_P)$ such that $N_P = \sigma_P M_P$. We say a lattice N on W is in the *class* $[M]$ of M if there exists an isometry σ in $U(M)$ such that $N = \sigma M$. It is known that the number of isometric classes in a genus is finite (see Rehmann [12], Satz 3, page 37).

We say the hermitian form h is *totally positive definite* if h is positive definite at every infinite prime spot P of K.

Henceforth we assume h is a totally positive definite hermitian form. Then for any lattice N in the genus G_M of M the unitary group $U(M)$ is a finite group. Define the *mass* of the hermitian S-lattice M on W by

$$w(M) = \sum_{[N] \subseteq G_M} \frac{1}{|U(N)|}$$

Now we assume that h is integral, i.e., $h(M, M) \subset S$. Let P be a finite prime ideal in K, and μ be a positive integer. Let $A_{P\mu}(M, M)$ be the number of S-linear maps $\sigma :$ $M \longrightarrow M$ which are distinct modulo $P^\mu M$ and satisfy $h(\sigma x, \sigma y) \equiv h(x, y) \bmod P^\mu S$ for every x and y in M. Let p be the prime number in $\mathbb{Z}$ such that $P \mid p$ and f_P be the residual degree $[R/P : \mathbb{Z}/p]$. Then it is known that the limit

$$\lim_{\mu \to \infty} p^{-f_P r^2 \mu} A_{P\mu}(M, M)$$

exists (see Rehmann [12], Hilfssatz 5.3).

Define the hermitian local density $\beta_P(M)$ of M by

$$\beta_P(M) = \lim_{\mu \to \infty} p^{-f_P r^2 \mu} A_{P\mu}(M, M).$$

Then we have the following mass formula of totally positive definite hermitian S-lattice M on W (see Rehmann [12], (4.5), also Braun [3, Satz VI] and Böge [2, pp. 112]).

$$(2) \qquad w(M) = 2 N_{K/\mathbb{Q}}(D(E/K))^{\frac{r(r+1)}{4}} \left(\prod_{j=1}^{r} \frac{(j-1)!}{(2\pi)^j}\right)^{\frac{\varphi(q)}{2}}$$

$$\cdot D(K/\mathbb{Q})^{\frac{r^2}{2}} N_{K/\mathbb{Q}}(\delta M)^r \prod_{P \text{ finite}} \beta_P(M)^{-1}.$$

Where $D(E/K)$ is the discriminant of the field E over K, $D(K/\mathbb{Q})$ is the absolute discriminant of the field K, $N_{K/\mathbb{Q}}(D(E/K)) = |R/D(E/K)|$, $N_{K/\mathbb{Q}}(\delta M) = |R/\delta M|$, and $\varphi(q)$ is the Euler number of q.

CHAPTER II
GENERAL THEORY

§3. Lattices with nontrivial automorphisms whose minimal polynomials are reducible.

Let M be a lattice in the genus G_L of L of type $R(q)$ with an isometry g of order prime $q \geq 2$. Let $G = \langle g \rangle$ be the cyclic group of order q generated by g. Let $\mathbb{Q}G$ be the group algebra and $\mathbb{Z}G$ be the group ring. Let ς be a primitive q-th root of unity. Then the group algebra $\mathbb{Q}G$ is isomorphic to $\mathbb{Q} \times \mathbb{Q}(\varsigma)$ under mapping which corresponds g to $(1, \varsigma)$. Let $\Gamma = \mathbb{Z} \times \mathbb{Z}[\varsigma] \subset \mathbb{Q} \times \mathbb{Q}(\varsigma)$. Then $\mathbb{Z}G$ is injected into Γ by above isomorphism. We often identify $\mathbb{Q}G$ and $\mathbb{Q} \times \mathbb{Q}(\varsigma)$ in the following arguments. The following arguments are based on Quebbemann [11], §2.

Let $W = \mathbb{Q}M$, $W_0 = (\sum_{i=0}^{q-1} g^i)W$ and $W_1 = (g - 1)W$. Then we can easily see that $W = W_0 \perp W_1$, $gx = x$ for every $x \in W_0$ and $(\sum_{i=0}^{q-1} g^i)x = 0$ for every $x \in W_1$. The group algebra $\mathbb{Q}G$ acts on W. With this action M is a $\mathbb{Z}G$-module. Let $h(x,y) = \sum_{i=0}^{q-1} B(g^{-i}x, y)g^i$ for x, y in W. Then h is an hermitian form with respect to the involution sending g to g^{-1}. Clearly $h(x,y) \in \mathbb{Z}G$ for any x and y in M. Let us denote the hermitian $\mathbb{Q}G$ structure of W (resp. $\mathbb{Z}G$ structure of M) by $\tilde{W}$ (resp. by $\tilde{M}$).

Proposition 3.1. Let $y \in \tilde{W}$, then $h(x,y) \in \mathbb{Z}G$ for any $x \in \tilde{M}$ if and only if $y \in \tilde{M}$.

Proof. Assume $h(x,y) \in \mathbb{Z}G$ for any $x \in \tilde{M}$. Then by the definition of $h(x,y)$ we have $B(x,y) \in \mathbb{Z}$ for any $x \in M$. Therefore, by 82 : 14b in [8] we have $y \in M = \tilde{M}$. Conversely, if $y \in \tilde{M} = M$ then $B(g^{-i}x, y) \in \mathbb{Z}$ for any $x \in M$ because $g \in O(M)$. Hence we have $h(x,y) \in \mathbb{Z}G$. $\qquad\square$

For any $x, y \in \tilde{W}$, let $h(x,y) = (B_0(x,y), h_1(x,y)) \in \mathbb{Q} \times \mathbb{Q}(\varsigma)$.

Proposition 3.2. Let $x, y \in W$.

(i)
$$B_0(x,y) = \sum_{i=0}^{q-1} B(g^{-i}x, y) \in \mathbb{Q},$$

$$h_1(x,y) = \sum_{i=0}^{q-1} B(g^{-i}x, y)\varsigma^i \in \mathbb{Q}(\varsigma).$$

11

(ii) Let $x = x_0 + x_1$, $y = y_0 + y_1$ where $x_i, y_i \in W_i$ for $i = 0, 1$. Then $h(x, y) = (B_0(x_0, y_0), h_1(x_1, y_1))$.

Proof. (i) is clear because $g^i \in \mathbb{Q}G$ corresponds to $(1, \varsigma^i) \in \mathbb{Q} \times \mathbb{Q}(\varsigma)$. As for (ii), we have from (i)

$$B_0(x, y) = \sum_{i=0}^{q-1} B(g^{-i}x, y) = \sum_{i=0}^{q-1} B(g^{-i}(x_0 + x_1), y_0 + y_1)$$

$$= B(\sum_{i=0}^{q-1} g^{-i}x_0 + \sum_{i=0}^{q-1} g^{-i}x_1, y_0 + y_1).$$

Since $x_0 \in W_0 = (\sum_{i=0}^{q-1} g^i)W$, we have $g^{-i}x_0 = x_0$ for $i = 0, \cdots, q - 1$ and therefore $(\sum_{i=0}^{q-1} g^{-i})x_0 = qx_0$. Since $x_1 \in W_1 = (g - 1)W$ we have $(\sum_{i=0}^{q-1} g^{-i})x_1 = 0$.

Therefore we have $B_0(x, y) = B(qx_0, y_0 + y_1) = B(qx_0, y_0) = B(\sum_{i=0}^{q-1} g^{-i}x_0, y_0) = B_0(x_0, y_0)$.

Similarly $h_1(x, y) = \sum_{i=0}^{q-1} B(g^{-i}x, y)\varsigma^i$

$$= \sum_{i=0}^{q-1} B(g^{-i}x_0 + g^{-i}x_1, y_0 + y_1)\varsigma^i$$

$$= \sum_{i=0}^{q-1} B(g^{-i}x_0, y_0 + y_1)\varsigma^i + \sum_{i=0}^{q-1} B(g^{-i}x_1, y_0 + y_1)\varsigma^i$$

$$= \sum_{i=0}^{q-1} B(x_0, y_0 + y_1)\varsigma^i + \sum_{i=0}^{q-1} B(g^{-i}x_1, y_0 + y_1)\varsigma^i$$

Since $\sum_{i=0}^{q-1} \varsigma^i = 0$ and $g^{-i}x_1 \in W_1$, we have $h_1(x, y) = \sum_{i=0}^{q-1} B(g^{-i}x_1, y_1)\varsigma^i = h_1(x_1, y_1)$. $\square$

The isomorphism between $\mathbb{Z}G$ and $\mathbb{Q} \times \mathbb{Q}(\varsigma)$ sends $\frac{1}{q}\sum_{i=0}^{q-1} g^i$ to $(1, 0)$ and $g - \frac{1}{q}(\sum_{i=0}^{q-1} g^i)$ to $(0, \varsigma)$. Moreover for any $x \in W_0$ we have $(\frac{1}{q}\sum_{i=0}^{q-1} g^i)x = x$ and for any $x \in W_1$, $[g - \frac{1}{q}(\sum_{i=0}^{q-1} g^i)]x = gx$.

Therefore we can consider W_0 as a vector space over $\mathbb{Q}$ and W_1 as a vector space over $\mathbb{Q}(\varsigma)$. Where, for any $x \in W_1, \varsigma \cdot x = gx$.

Proposition 3.3. (i) Restrict B_0 to W_0, then (W_0, B_0) is a positive definite quadratic space over $\mathbb{Q}$. Moreover $B_0(x, y) = qB(x, y)$ for any $x, y \in W_0$

(ii) Restrict h_1 to W_1. If $q \neq 2$ then (W_1, h_1) is a totally positive definite hermitian space over $\mathbb{Q}(\varsigma)$ with respect to complex conjugation. Moreover we have $Tr_{\mathbb{Q}(\varsigma)/\mathbb{Q}}(h_1(x, y)) = qB(x, y)$ for any $x, y \in W_1$. If $q = 2$, then (W_1, h_1) is also a positive definite quadratic space and $h_1(x, y) = 2B(x, y)$ for any $x, y \in W_1$.

Proof (i) As we showed in the proof of Proposition 3.2 $B_0(x, y) = qB(x, y)$ for all $x, y \in W_0$, (i) is clear.

(ii) We have $\overline{\tilde{h_1}(y, x)} = \sum_{i=0}^{q-1} \overline{B(g^{-i}y, x)\varsigma^i}$

$$= \sum_{i=0}^{q-1} B(g^{-i}y, x)\overline{\varsigma}^i = \sum_{i=0}^{q-1} B(y, g^i x)\varsigma^{-i}$$

$$= \sum_{i=0}^{q-1} B(g^i x, y)\varsigma^{-i} = \sum_{i=0}^{q-1} B(g^{-i}x, y)\varsigma^i$$

$$= h_1(x, y)$$

and $h_1(x, \varsigma \cdot y) = \sum_{i=0}^{q-1} B(g^{-i}x, gy)\varsigma^i$

$$= \sum_{i=0}^{q-1} B(q^{-i-1}x, y)\varsigma^i = \sum_{i=0}^{q-1} B(g^{-(i+1)}x, y)\varsigma^{i+1-1}$$

$$= \varsigma^{-1} \sum_{i=0}^{q-1} B(g^{-(i+1)}x, y)\varsigma^{i+1} = \varsigma^{-1}h_1(x, y)$$

$$= \overline{\varsigma}h_1(x, y).$$

Therefore $h_1(x, y)$ is a hermitian form with respect to complex conjugation. Moreover we have $Tr_{\mathbb{Q}(\varsigma)/\mathbb{Q}}(h_1(x, y))$

$$= Tr_{\mathbb{Q}(\varsigma)/\mathbb{Q}}(\sum_{i=0}^{q-1} B(g^{-i}x, y)\varsigma^i) = (q - 1)B(x, y)$$

$$- \sum_{i=1}^{q-1} B(g^{-i}x, y) = qB(x, y) - \sum_{i=0}^{q-1} B(g^{-i}x, y)$$

$$= qB(x, y) - B(\sum_{i=0}^{q-1} g^{-i}x, y) = qB(x, y).$$

Since (W_1, B) is a positive definite quadratic space over $\mathbb{Q}$, (W_1, h_1) is a totally positive definite hermitian space (see Biermann ([1], pp. 47, Folgerung 2) also Feit ([5], Theorem 6.1) □

Let $\Gamma \tilde{M} = M_0 \times M_1$ where $M_0 = (\mathbb{Z} \times \{0\})\tilde{M}$ and $M_1 = (\{0\} \times \mathbb{Z}[\varsigma])\tilde{M}$. Then M_i is a lattice on W_i for $i = 0, 1$. For every x and y in $\Gamma \tilde{M}, h(x, y) \in \Gamma$. Therefore $B_0(x, y) \in \mathbb{Z}$ and $h_1(x, y) \in \mathbb{Z}[\varsigma]$ for any $x, y \in \Gamma \tilde{M}$. Hence (M_0, B_0) is a positive definite integral $\mathbb{Z}$-lattice on W_0 and (M_1, h_1) is a totally positive definite hermitian $\mathbb{Z}[\varsigma]$-lattice on W_1. Let us denote (M_1, h_1) by $(\mathcal{M}_1, h_1)$. If $q = 2$ then $\mathcal{M}_1$ is also a positive definite integral $\mathbb{Z}$-lattice. In this case we denote (M_1, h_1) by $(\mathcal{M}_1, B_1)$.

Proposition 3.4. (i) Let $y \in W_0$ and $B_0(x, y) \in q\mathbb{Z}$ for any x in M_0, then $y \in M_0$.

(ii) Let $y \in W_1$ and λ be a prime element in $\mathbb{Z}[\varsigma]$ above q. Assume $h_1(x, y) \in \lambda\mathbb{Z}[\varsigma]$ for every $x \in \mathcal{M}_1$, then $y \in \mathcal{M}_1$. Here if $q = 2$, then $\varsigma = -1$ and $\lambda = 2$.

Proof (i) Since x and y are in $W_0, B_0(x, y) = qB(x, y)$. Therefore we have $B(x, y) \in \mathbb{Z}$ for all $x \in M_0$. Since W_1 and W_0 are orthogonal to each other with respect to $B, B(x, y) \in \mathbb{Z}$ for all $x \in \Gamma\tilde{M}$ and hence for all $x \in M$. Therefore $y \in M \cap W_0 \subset M_0$.

(ii) Since $Tr_{\mathbb{Q}(\varsigma)/\mathbb{Q}}(\lambda) = \pm q$, we have $qB(x, y) = Tr_{\mathbb{Q}(\varsigma)/\mathbb{Q}}(h_1(x, y)) \in Tr_{\mathbb{Q}(\varsigma)/\mathbb{Q}}(\lambda\mathbb{Z}[\varsigma]) \subset q\mathbb{Z}$. Hence we have $B(x, y) \in \mathbb{Z}$ for all $x \in \mathcal{M}_1$. A similar argument as in (i) gives $y \in \mathcal{M}_1$.   □

Thus, we constructed an integral lattice $\mathcal{M}_0$ and a hermitian $\mathbb{Z}[\varsigma]$-lattice $\mathcal{M}_1$. The construction depends on the choice of g. Therefore, it is possible to construct more than one pair $(\mathcal{M}_0, \mathcal{M}_1)$ of lattices from one class in G_L. It is also possible to obtain the same pair $(\mathcal{M}_0, \mathcal{M}_1)$ from two non-isomorphis lattices. However Lemma 3.10 and 3.13 in this section will show it does not matter so much.

Let $\Lambda = \mathbb{Z}G$ and $I = q\Lambda + (1 - g)\Lambda \subset \mathbb{Z}G$. Then I corresponds to $q\mathbb{Z} \times \lambda\mathbb{Z}[\varsigma]$ in $\Gamma = \mathbb{Z} \times \mathbb{Z}[\varsigma]$. Let $V = \Gamma\tilde{M}/I\tilde{M}, V_0 = M_0/qM_0$ and $V_1 = \mathcal{M}_1/\lambda\mathcal{M}_1$. Then $V = V_0 \times V_1$. Define bilinear forms b, b_0 and b_1 on V, V_0 and V_1 induced by h, B_0 and h_1 respectively, i.e. $b(\bar{x}, \bar{y}) = h(x, y) \, mod \, I, b_0(\bar{x}_0, \bar{y}_0) = B_0(x_0, y_0) \, mod \, q\mathbb{Z}$ and $b_1(\bar{x}_1, \bar{y}_1) = h_1(x_1, y_1) \, mod \, \lambda\mathbb{Z}[\varsigma]$ where x and y are in $\Gamma\tilde{M}$, x_0 and y_0 are in M_0, x_1 and y_1 are in $\mathcal{M}_1$, and $\bar{x}$ is the equivalent class of x. The quotient $\Lambda/I \cong \mathbb{F}_q$ (finite field of q elements) is injected onto the diagonal Δ in $\Gamma/I \cong \mathbb{F}_q \times \mathbb{F}_q$.

Let $\mathcal{L}_{\tilde{M}}$ be the set of all hermitian Λ-lattices $\tilde{K}$ in $\Gamma\tilde{M}$ which satisfy the following two conditions :

(i) $\Gamma\tilde{K} = \Gamma\tilde{M}$.

(ii) For any given $y \in \Gamma\tilde{M}$ the necessary and sufficient condition that $h(x, y) \in \Lambda$ for all $x \in \tilde{K}$ is $y \in \tilde{K}$. Here $\tilde{K}$ may not necessarily be constructed from a lattice in the genus G_L of L.

Let $\tilde{K}$ be in $\mathcal{L}_{\tilde{M}}, U_{\tilde{K}} = \tilde{K}/I\tilde{K}, U_{\tilde{K}}^{(0)} = U_{\tilde{K}} \cap (V_0 \times \{0\})$ and $U_{\tilde{K}}^{(1)} = U_{\tilde{K}} \cap (\{0\} \times V_1)$. Then we have the following Proposition 3.5.

This Proposition 3.5 is very crucial in our whole argument and is proved by Quebbmann (Lemma 2.2 in [11]), but for the purpose of making this thesis self contained and easier to read, we will give the proof here.

Proposition 3.5 (Quebbemann [11, Lemma (2.2)])

(i) $U_{\tilde{K}}^{(0)} \oplus U_{\tilde{K}}^{(1)}$ is a maximal Γ/I submodule of V contained in $U_{\tilde{K}}$.

(ii) $V^{\perp} = U_{\tilde{K}}^{(0)} \oplus U_{\tilde{K}}^{(1)}$.

(iii) Let $V/V^{\perp} = V_0' \times V_1'$, then there exists an isometry $\tau_{\tilde{K}} : V_0' \cong V_1'$ such that $U_{\tilde{K}}/V^{\perp} = \{(v_0, \tau_{\tilde{K}}(v_0)) \mid v_0 \in V_0'\}$. Therefore $U_{\tilde{K}} = \{(v_0, \tau_{\tilde{K}}(v_0)) \mid v_0 \in V_0'\} \oplus V^{\perp}$ because $V_0' \times V_1'$ can be considered as a direct summand of V.

Proof (i) Let $u = (u_0, 0)$ and $v = (v_0, 0)$ be in $U_{\tilde{K}}^{(0)}$. Since $\tilde{K}$ is a Λ-lattice $u + v \in U_{\tilde{K}}$. Therefore $u + v \in U_{\tilde{K}}^{(0)}$. Let $a = (a_0, a_1) \in \mathbb{F}_q \times \mathbb{F}_q$, then $au = (a_0 u_0, 0) = (a_0, a_0)(u_0, 0)$. Since $(a_0, a_0) \in \Lambda/I$ we have $au \in U_{\tilde{K}}^{(0)}$. Therefore $U_{\tilde{K}}^{(0)}$ is an $\mathbb{F}_q \times \mathbb{F}_q$ module. A similar argument shows $U_{\tilde{K}}^{(1)}$ is also an $\mathbb{F}_q \times \mathbb{F}_q$ module. Let U' be an $\mathbb{F}_q \times \mathbb{F}_q$ module with $U_{\tilde{K}}^{(0)} + U_{\tilde{K}}^{(1)} \subseteq U' \subseteq U_{\tilde{K}}$. Let $u = (u_0, u_1) \in U'$. Since U' is an $\mathbb{F}_q \times \mathbb{F}_q$-submodule of V, $(1, 0)u$ and $(0, 1)u$ are in U'. Therefore $(1, 0)u \in U_{\tilde{K}}^{(0)}$ and $(0, 1)u \in U_{\tilde{K}}^{(1)}$ and hence we have $u = (1, 0)u + (0, 1)u \in U_{\tilde{K}}^{(0)} \oplus U_{\tilde{K}}^{(1)}$. This shows $U' = U_{\tilde{K}}^{(0)} \oplus U_{\tilde{K}}^{(1)}$.

(ii) Since $\Gamma \tilde{K} = \Gamma \tilde{M}$, $U_{\tilde{K}}$ generates V over Γ/I. Therefore it is sufficient to show that $U_{\tilde{K}}^{\perp} = U_{\tilde{K}}^{(0)} \oplus U_{\tilde{K}}^{(1)}$. Let $u \in U_{\tilde{K}}^{\perp}$, then $b(v, u) = 0$ for all $v \in U_{\tilde{K}}$, i.e., $b(v, u) \in \Lambda/I$ for all v in $U_{\tilde{K}} = \tilde{K}/I\tilde{M}$. Since $\tilde{K}$ is in $\mathcal{L}_{\tilde{M}}$, this implies $u \in U_{\tilde{K}}$, that is, $U_{\tilde{K}}^{\perp} \subset U_{\tilde{K}}$. Hence by (i) of this proposition it is enough to show $U_{\tilde{K}}^{(0)} \oplus U_{\tilde{K}}^{(1)} \subseteq U_{\tilde{K}}^{\perp}$. Let $u = (u_0, 0) \in U_{\tilde{K}}^{(0)}$ and $v = (v_0, v_1) \in U_{\tilde{K}}$, then $b(u, v) = (b_0(u_0, v_0), b_1(0, v_1)) = (b_0(u_0, v_0), 0)$. On the other hand we have $b(u, v) \in \Delta$. Therefore $b_0(u_0, v_0) = 0$. This means $b(u, v) = 0$, i.e., $U_{\tilde{K}}^{(0)} \subseteq U_{\tilde{K}}^{\perp}$. Similarly we can show $U_{\tilde{K}}^{(1)} \subseteq U_{\tilde{K}}^{\perp}$. This gives $U_{K}^{(0)} \oplus U_{K}^{(1)} \subseteq U_{\tilde{K}}^{\perp}$.

(iii) Let $u = (u_0, u_1)$ and $v = (v_0, v_1) \in U_{\tilde{K}}$. If $(u_0 - v_0, 0) \in U_{\tilde{K}}^{(0)}$, then $(1, 0)(u - v) = (u_0 - v_0, 0) \in U_{\tilde{K}}^{(0)} \subseteq U_{\tilde{k}}$. Since $U_{\tilde{K}}$ is a Λ/I module, we have $(1, 1)(u - v) \in U_{\tilde{K}}$. Therefore we have $(0, u_1 - v_1) = (0, 1)(u - v) = (1, 1)(u - v) - (1, 0)(u - v) \in U_{\tilde{K}}$. Hence $(0, u_1 - v_1) \in U_{\tilde{K}}^{(1)}$. This means for any u and v in $U_{\tilde{K}}$ if $u_0 \equiv v_0 \mod U_{\tilde{K}}^{(0)}$ then $u \equiv v \mod U_{\tilde{K}}^{(0)} \oplus U_{\tilde{K}}^{(1)}$, that is the second component of $U_{\tilde{K}} \mod U_{\tilde{K}}^{(0)} \oplus U_{\tilde{K}}^{(1)}$ corresponds to the first component uniquely. Let us denote this correspondence with $\tau_{\tilde{K}}$. Let us assume that $\bar{u}_0 \in V_0'$ and $(\bar{u}_0, \tau_{\tilde{K}}(\bar{u}_0))$ is in $U_{\tilde{K}}/U_{\tilde{K}}^{(0)} \oplus U_{\tilde{K}}^{(1)}$. Since $(a, a)(\bar{u}_0, \tau_{\tilde{K}}(\bar{u}_0)) = (a\bar{u}_0, a\tau_{\tilde{K}}(\bar{u}_0)) \in U_{\tilde{K}}/U_{\tilde{K}}^{(0)} \oplus U_{\tilde{K}}^{(1)}$, we have $\tau_{\tilde{K}}(a\bar{u}_0) = a\tau_{\tilde{K}}(\bar{u}_0)$. Since $U_{\tilde{K}}$ generates V over Γ/I we can extend $\tau_{\tilde{K}}$ linearly to V_0' as follows. For any $v_0 \in V_0$ let $(v_0, 0) = \sum_i (a_{i0}, a_{i1})(u_{i0}, u_{i1})$ where $a_{i0}, a_{i1} \in \mathbb{F}_q$ and $(u_{i0}, u_{i1}) \in U_{\tilde{K}}$. Since $\sum_i (a_{i0}, a_{i1})(u_{i0}, u_{i1}) = (\sum_i a_{i0} u_{i0}, \sum_i a_{i1} u_{i1})$ we have $\sum_i a_{i1} u_{i1} = 0$. Hence we have $(v_0, 0) = (\sum_i a_{i0} u_{i0}, 0)$ for any $v_0 \in V_0$. Define $\tau_{\tilde{K}}(\bar{v}_0) = \sum_i a_{i0} \tau_{\tilde{K}}(\bar{u}_{i0})$. Clearly $\tau_{\tilde{K}}$ is an injection from V_0' to V_1'. We will show that $\tau_{\tilde{K}}$ is surjective. Let $v_1 \in V_1$. Then $(0, \bar{v}_1) = \sum_i (a_{i0}, a_{i1})(\bar{u}_{i0}, \bar{u}_{i1}) = (\sum_i a_{i0} \bar{u}_{i0}, \sum_i a_{i1} \bar{u}_{i1}) = (0, \sum_i a_{i1} \bar{u}_{i1})$ with some $(a_{i0}, a_{i1}) \in \mathbb{F}_q \times \mathbb{F}_q$ and $(u_{i0}, u_{i1}) \in U_{\tilde{K}}$. Then $(\bar{u}_{i0}, \bar{u}_{i1}) = (\bar{u}_{i0}, \tau_{\tilde{K}}(\bar{u}_{i0}))$. Therefore we have $\bar{v}_1 = \sum_i a_{i1} \bar{u}_{i1} = \sum_i a_{i1} \tau_{\tilde{K}}(\bar{u}_{i0}) = \tau_{\tilde{K}}(\sum_i a_{i1} \bar{u}_{i0})$. For any u and v in $U_{\tilde{K}}, b(u, v) \in \Lambda I = \Delta$, we have $b_0(u_0, v_0) = b_1(u_1, v_1)$, this shows that $\tau_{\tilde{K}}$ is an isometry.

Remark. By Proposition 3.1 we have $\tilde{M} \in \mathcal{L}_{\tilde{M}}$, therefore V_0' is isometric to V_1'. Let $I(V_0', V_1')$ be the set of all the isometries from V_0' to V_1'. Then we have the following proposition.

Proposition 3.6. Let τ be an element in $I(V_0', V_1')$. Let π be the projection $\Gamma\tilde{M} \longrightarrow \Gamma\tilde{M}/I\tilde{M}$, and $U = \{(v_0, \tau(v_0)) \mid v_0 \in V_0'\} \oplus V^{\perp}$. Then $\tilde{K} = \pi^{-1}(U)$ is in $\mathcal{L}_{\tilde{M}}$.

Proof. V_0 is a finite dimensional $\mathbb{F}_q$-vector space, and $v = (v_0, v_1)$ in $V_0' \times V_1'$ can be written as $v_0 = \sum_i a_i u_i$ and $\sum_i b_i \tau(u_i)$ with some $a_i, b_i \in \mathbb{F}_q$ and $u_i \in V_0'$. Therefore $v = (\sum_i a_i u_i, \sum_i b_i \tau(u_i)) = \sum_i (a_i, b_i)(u_i, \tau(u_i)) \in (\Gamma/I)U$. This proves $\Gamma\tilde{K} = \Gamma\tilde{M}$. Next we will show that for any $y \in \Gamma\tilde{M}$ the necessary and sufficient condition that $h(x, y) \in \Lambda$ for all $x \in \tilde{K}$ is $y \in \tilde{K}$. Suppose $h(x, y) \in \Lambda$ for any x in $\tilde{K}$. Let $\bar{x} = (\bar{x}_0, \tau(\bar{x}_0)) \oplus \bar{x}'$ and $\bar{y} = (\bar{y}_0, \bar{y}_1) \oplus \bar{y}'$ where $\bar{x}_0 \in V_0'$, $(\bar{y}_0, \bar{y}_1) \in V_0' \times V_1'$ and $\bar{x}', \bar{y}' \in V^{\perp}$. Then $b(\bar{x}, \bar{y}) = (b_0(\bar{x}_0, \bar{y}_0), b_1(\tau(\bar{x}_0), \bar{y}_1)) \in \Delta$. Therefore $b_0(\bar{x}_0, \bar{y}_0) = b_1(\tau(\bar{x}_0), \bar{y}_1)$. On the other hand, we have $b_1(\tau(\bar{x}_0), \tau(\bar{y}_0)) = b_0(\bar{x}_0, \bar{y}_0)$. Therefore $b_1(\tau(\bar{x}_0), \bar{y}_1 - \tau(\bar{y}_0)) = 0$ for all $\bar{x}_0 \in V_0'$ and hence we have $\bar{y}_1 = \tau(\bar{y}_0)$. This shows $y \in \tilde{K}$. □

Proposition 3.7. There is a one to one correspondence between $\mathcal{L}_{\tilde{M}}$ and $I(V_0', V_1')$.

Proof. This is clear from Propositions 3.5 and 3.6. □

Proposition 3.8. Let $\tilde{K}$ and $\tilde{N}$ be in $\mathcal{L}_{\tilde{M}}$. Assume that there exists an isometry $\sigma = (\sigma_0, \sigma_1)$ in $Aut\,M_0 \times Aut\,\mathcal{M}_1$ which induces the isometry from $U_{\tilde{K}} = \tilde{K}/I\tilde{M}$ to $U_{\tilde{N}} = \tilde{N}/I\tilde{M}$. Then $\tilde{K}$ and $\tilde{N}$ are isometric as hermitian Λ-lattices.

Proof. This is clear because $\sigma(\tilde{K}) = \sigma(\tilde{K}+I\tilde{K}) = \sigma(\tilde{K}+I\tilde{M}) = \tilde{N}+I\tilde{M} = \tilde{N}+I\tilde{N} = \tilde{N}$.

Proposition 3.9.

$$\sum_{Class\,\tilde{K} \in \mathcal{L}_{\tilde{M}}} \frac{1}{|Aut(\tilde{K})|} \leq \frac{|I(V_0', V_1')|}{|Aut\,M_0||Aut\,\mathcal{M}_1|}$$

Here $Aut\,M_0$ is the orthogonal group of M_0 and $Aut\,\mathcal{M}_1$ is the unitary group of $\mathcal{M}_1$. If $q = 2$ then $Aut\,\mathcal{M}_1$ is also the orthogonal group of $\mathcal{M}_1 = M_1$.

Proof. Let $\tilde{K} \in \mathcal{L}_{\tilde{M}}$ and $\sigma = \sigma_0 \times \sigma_1 \in Aut\,M_0 \times Aut\,\mathcal{M}_1$. Clearly we have $\sigma\tilde{K} \in \mathcal{L}_{\tilde{M}}$. Let $H = Aut\,M_0 \times Aut\,\mathcal{M}_1$. Let $\mathcal{L}_{\tilde{M}} = H\tilde{K}_1 \cup \cdots \cup H\tilde{K}_s$ be the decomposition of $\mathcal{L}_{\tilde{M}}$ into the orbits given by the action of H on $\mathcal{L}_{\tilde{M}}$. Let $H_i = \{\sigma \in H \mid \sigma\tilde{K}_i = \tilde{K}_i\}$ and ℓ_i be the length of the orbit of $\tilde{K}_i$, that is, $\ell_i = |\{\tilde{K} \in \mathcal{L}_{\tilde{M}} | \tilde{K} = \sigma(\tilde{K}_i) \text{ with some } \sigma \in H\}|$. Then for any $i = 1, \cdots, s$ we have $|H| = \ell_i|H_i|$. Therefore we have

$$\sum_{i=1}^{s} \frac{1}{|H_i|} = \sum_{i=1}^{s} \frac{\ell i}{|H|} = \frac{|\mathcal{L}_{\tilde{M}}|}{|H|} = \frac{|I(V_0', V_1')|}{|H|}.$$

By Proposition 3.8, the Λ-lattices in the orbit $H\tilde{K}_i$ are isometric for each $i = 1, \cdots, s$. On the other hand, H_i is a subgroup of $Aut\,(\tilde{K}_i)$. Therefore we have

$$\sum_{Class\,\tilde{K}\,in\,\mathcal{L}_{\tilde{M}}} \frac{1}{|Aut\,\tilde{K}|} \leq \sum_{i=1}^{s} \frac{1}{|Aut\,\tilde{K}_i|} \leq \sum_{i=1}^{s} \frac{1}{|H_i|}.$$

This gives the Proposition 3.9. □

Lemma 3.10. Let M be in the genus G_L of L and of type $R(q)$. Then we have the following inequality.

$$\sum_{\substack{cls\,K\subseteq G_L \\ of\,type\,R(q) \\ with\,\Gamma\tilde{K}\cong\Gamma\tilde{M}}} \frac{1}{|O(K)|} \leq \frac{|I(V_0', V_1')|}{|Aut\,M_0||Aut\,M_1|}$$

Proof. Let K and N be the classes in G_L of type $R(q)$ with $\Gamma\tilde{K} \cong \Gamma\tilde{N} \cong \Gamma\tilde{M}$. Then it is clear from the construction of $\tilde{K}$ and $\tilde{N}$ that a Λ-isometry $\sigma : (\tilde{K}, h_K) \longrightarrow (\tilde{N}, h_N)$ induces an $\mathbb{Z}$-isometry $\sigma : (K, B) \longrightarrow (N, B)$. Therefore we have

$$\sum_{\substack{cls\,K\subseteq G_L \\ of\,type\,R(q) \\ with\,\Gamma\tilde{K}\cong\Gamma\tilde{M}}} \frac{1}{|O(K)|} \leq \sum_{\substack{class\,\tilde{K} \\ constructed \\ from\,K\in G_L \\ with\,\Gamma\tilde{K}\cong\Gamma\tilde{M}}} \frac{1}{|Aut\,(\tilde{K})|} \leq \sum_{\substack{Class\,\tilde{K} \\ in\,\mathcal{L}_{\tilde{M}}}} \frac{1}{|Aut\,(\tilde{K})|}$$

Hence by Proposition 3.9 we have the proof for Lemma 3.10. □

Let $L(q, r, \rho)$ and $G(q, r, \rho)$ be the set defined in the introduction. Then there are finitely many representatives of lattices $M^{(1)}, \cdots, M^{(k)}$ in $L(q, r, \rho)$ with the following conditions:

(i) $(\Gamma\tilde{M}^{(i)}, h^{(i)})$ is not isometric to $(\Gamma\tilde{M}^{(j)}, h^{(j)})$ for $i \neq j$ as Γ-lattices.

(ii) For every $K \in L(q, r, \rho)$ there exists some $M^{(i)}$ such that $(\Gamma\tilde{K}, h_K) \cong (\Gamma\tilde{M}^{(i)}, h^{(i)})$. Lemma 3.10 gives the following proposition.

Proposition 3.11.

$$\sum_{cls\,K\subseteq L(q,r,\rho)} \frac{1}{|O(K)|} \leq \sum_{i=1}^{k} \frac{|I(V_0'(M_0^{(i)}), V_1'(M_1^{(i)}))|}{|Aut\,M_0^{(i)}||Aut\,M_1^{(i)}|}$$

Proposition 3.12. Let M be a lattice in $L(q, r, \rho)$. Then the pair of genera (G_{M_0}, G_{M_1}) of the lattices M_0 and M_1 which are constructed from M is contained in $G(q, r, \rho)$.

Proof. Since $(1, 0)$ corresponds to $\frac{1}{q}\sum_{i=0}^{q-1} g^i$ in $\mathbb{Q}G$ and $(0, 1)$ corresponds to $1 - \frac{1}{q}(\sum_{i=0}^{q-1} g^i)$ in $\mathbb{Q}G$, we have $M_0 = (\frac{1}{q}\sum_{i=0}^{q-1} g^i)M$ and $M_1 = (1 - \frac{1}{q}(\sum_{i=0}^{q-1} g^i))M$. Then M_0 and M_1 are lattice on W_0 and W_1 respectively. Therefore we have $q(M_0 \perp M_1) \subseteq M \subseteq M_0 \perp M_1$ with respect to the bilinear form B. This shows that the discriminant of $qM_i (i = 0, 1)$ with respect to the bilinear form B is power of q. On the other hand by Proposition 3.3, we have $B_0(x, y) = qB(x, y)$ for any $x, y \in M_0 = M_0$ (if $q = 2$ $B_1(x, y) = 2 B(x, y)$ for any $x, y \in M_1 = M_1$). Hence discriminant of M_0 with respect to the bilinear form B_0 is also a power of q (if $q = 2$ the discriminant of M_1 with respect to the bilinear form B_1 is also power of 2).

Proposition 3.4 shows that $(M_0)_p$ is unimodular at the prime $p \neq q$ and $(M_0)_q$ has a Jordan splitting (see O'Meara [8] pp. 243) $J_0 \perp J_1$ with some unimodular $\mathbf{Z}_q$ lattice J_0 and $q\mathbf{Z}_q$-modular $\mathbf{Z}_q$-lattice J_1 ($(M_0)_q$ could be unimodular or $q\mathbf{Z}_q$-modular). If $q = 2$ then this is also true for $(M_1)_2$. Since $dim_{\mathbb{F}_q} V_0'(M_0) = dim_{\mathbb{F}_q} V_1'(M_1) = \rho$, the rank of J_0 has to be ρ. This shows that $d_{B_0} M_0 = q^{m_0 - \rho}$ (if $q = 2$ then $d_{B_1} M_1 = q^{r-\rho}$). Similarly Proposition 3.4 shows that $(M_1)_P$ is unimodular at the prime $P \nmid q$ and $(M_1)_P$ $P \mid q$ has a Jordan splitting $J_0 \perp J_1$ with some unimodular hermitian S_P-lattice J_0 and λS_P-modular hermitian S_P-lattice J_1 (see Jacobowitz [7] page 448, 449). Since $dim_{\mathbb{F}_q} V_1'(M_1) = \rho$, the rank of J_0 has to be ρ. This shows that $N_{K/\mathbb{Q}}(\delta M_1) = q^{r-\rho}$. If $q = 2$ and a lattice M in G_L of type $R(2)$ has an isometry g which gives rank $_{\mathbf{Z}} M_0 < r$, then taking $-g$ we can consider $M \in L(q, r, \rho)$. The other conditions in the definition of $G(q, r, \rho)$ are clearly obtained from the conditions in the definition of $L(q, r, \rho)$.

Lemma 3.13.

(i) $\displaystyle\sum_{cls\, K \subseteq L(q,r,\rho)} \frac{1}{|O(K)|} \leq \sum_{(G_{N_0}, G_{N_1}) \in G(q,r,\rho)} |I(V_0'(N_0), V_1'(N_1)| \omega(N_0)\omega(N_1).$

(ii) $\displaystyle \omega_{R(q)} < \sum_{r=1}^{[\frac{m-1}{q-1}]} \sum_{\rho=0}^{min(r,m_0)} \sum_{(G_{N_0}, G_{N_1}) \in G(q,r,\rho)} |I(V_0'(N_0), V_1'(N_1)| \omega(N_0)\omega(N_1).$

Proof. This is clear from the definition of mass, Propositions 3.11, and 3.12.

§4. Lattices with nontrivial automorphisms whose minimal polynomials are irreducible.

Case q = odd prime

Let M be a lattice in the genus G_L of type $IR(q)$ with the isometry g of order an odd prime number q. Let $W = \mathbb{Q}M$ and define the action of $\mathbb{Q}(\varsigma)$ on W through $\varsigma x = g(x)$ for $x \in W$. Then M is a $\mathbf{Z}[\varsigma]$-lattice with this action. Let $h(x, y) = \sum_{i=0}^{q-1} B(g^{-i}x, y)\varsigma^i$ for $x, y \in W$. Then $h(x, y)$ is a totally positive definite hermitian form on W with respect to complex conjugation. Clearly $h(x, y) \in \mathbf{Z}[\varsigma]$ for any $x, y \in M$. Thus W has a hermitian vector space structure over $\mathbb{Q}(\varsigma)$ and M has a hermititan $\mathbf{Z}[\varsigma]$-lattice structure. Let us denote them by $\mathcal{W}$ and $\mathcal{M}$, respectively.

Proposition 4.1. Let $y \in \mathcal{W}$. Then $h(x, y) \in \lambda\mathbf{Z}[\varsigma]$ for every $x \in \mathcal{M}$ if and only if $y \in \mathcal{M}$.

Proof. First assume $h(x,y) \in \lambda\mathbb{Z}[\varsigma]$ for every $x \in M$. Then by Proposition 3.3 (ii) we have $qB(x,y) = Tr_{\mathbb{Q}(\varsigma)/\mathbb{Q}}(h(x,y))$. Therefore by the assumption $qB(x,y) \in Tr_{\mathbb{Q}(\varsigma)/\mathbb{Q}}(\lambda\mathbb{Z}[\varsigma]) \subseteq q\mathbb{Z}$. Hence we have $B(x,y) \in \mathbb{Z}$ for every $x \in M = \mathcal{M}$. Since $\mathcal{M}$ is a unimodular lattice with respect to B, we have $y \in M = \mathcal{M}$. Conversely if $y \in \mathcal{M} = M$ then $B(x,y) \in \mathbb{Z}$ for all $x \in M = \mathcal{M}$. Therefore $Tr_{\mathbb{Q}(\varsigma)/\mathbb{Q}}(h(x,y)) = qB(x,y) \in q\mathbb{Z}$. This implies $h(x,y) \in \lambda\mathbb{Z}[\varsigma]$. □

Proposition 4.2. Let M and N be lattices in G_L of type $IR(q)$. Let $\mathcal{M}$ and $\mathcal{N}$ be the hermitian $\mathbb{Z}[\varsigma]$-lattices constructed from M and N, and let h_M and h_N be the hermitian form of $\mathcal{M}$ and $\mathcal{N}$, respectively. Suppose $(\mathcal{M}, h_M)$ and $(\mathcal{N}, h_N)$ are isometric as hermitian $\mathbb{Z}[\varsigma]$-lattices. Then (M, B) and (N, B) are isometric as $\mathbb{Z}$-lattices.

Proof. Let σ be the isometry from $\mathcal{M}$ to $\mathcal{N}$. Let g_M and g_N be the isometries of M and N which determine the structure $\mathcal{M}$ and $\mathcal{N}$, respectivley. Since $\sigma(\varsigma x) = \varsigma\sigma(x)$, $\varsigma \cdot x = g_M(x)$ for $x \in \mathcal{M}$ and $\varsigma \cdot u = g_N(u)$ for $u \in \mathcal{N}$, we have $\sigma g_M = g_N\sigma$. Hence we have
$$h_N(\sigma x, \sigma y) = \sum_{i=0}^{q-1} B(g_N^{-i}\sigma x, \sigma y)\varsigma^i$$

$$= \sum_{i=0}^{q-1} B(g_N^{-i}\sigma x, \sigma y)\varsigma^i - \sum_{i=0}^{q-1} B(g_N^{-(q-1)}\sigma x, \sigma y)\varsigma^i$$

$$= \sum_{i=0}^{q-2} \{B(g_N^{-i}\sigma x, \sigma y) - B(g_N^{-(q-1)}\sigma x, \sigma y)\}\varsigma^i$$

$$= \sum_{i=0}^{q-2} B((g_N^{-i} - g_N)\sigma x, \sigma y)\varsigma^i = \sum_{i=0}^{q-2} B(\sigma(g_M^{-i} - g_M)x, \sigma y)\varsigma^i.$$

On the other hand, we have $h_M(x,y) = \sum_{i=0}^{q-1} B(g_M^{-i}x, y)\varsigma^i$

$$= \sum_{i=0}^{q-2} B((g_M^{-i} - g_M^{-(q-1)})x, y)\varsigma^i = \sum_{i=0}^{q-2} B((g_M^{-i} - g_M)x, y)\varsigma^i.$$

Therefore we have $B(\sigma(1 - g_M)x, \sigma y) = B((1 - g_M)x, y)$ for all $x, y \in M = \mathcal{M}$.

The minimal polynomial of g_M is $x^{q-1} + \cdots + x + 1$, $1 - g_M$ is an automorphism of $W = \mathbb{Q}M$ as a vector space. This implies that $B(\sigma x, \sigma y) = B(x,y)$ for all $x, y \in W$. Hence σ is an isometry as $\mathbb{Z}$-lattices. □

Corollary 4.3.
$$|U(\mathcal{M})| \leq |O(M)|$$

Proof. Take $N = M$ in the proof of Proposition 4.2. □

Lemma 4.4. There is exactly one genus $G_\mathcal{M}$ of totally positive deifnite $\lambda\mathbb{Z}[\varsigma]$-modular hermitian $\mathbb{Z}[\varsigma]$-lattices of rank $\frac{m}{q-1}$ and we have

$$\omega_{IR(q)} \leq \omega(\mathcal{M}).$$

Proof. Proposition 4.1 and a similar argument as in [8], §82 F and §82G show that a hermitian $\mathbb{Z}[\varsigma]$-lattice $\mathcal{M}$, which is constructed from a integral $\mathbb{Z}$-lattice M in G_L of type $IR(q)$, is $\lambda\mathbb{Z}[\varsigma]$-modular. Proposition 3.2 in [14], Theorem 7.1 and Proposition 8.1 in [7] shows that there is exactly one genus of totally positive definite $\lambda\mathbb{Z}[\varsigma]$-modular hermitian $\mathbb{Z}[\varsigma]$-lattice of rank $\frac{m}{q-1}$. Therefore by Proposition 4.1 and Corollary 4.3, we have this lemma.   □

Case $q = 4$.

Let M be a lattice in G_L of type $IR(4)$ with the isometry g of order 4 whose minimal polynomial is $x^2 + 1$. Let $\varsigma = \sqrt{-1}, W = \mathbb{Q}M$. Let $\mathbb{Q}(\varsigma)$ act on M by $\varsigma \cdot x = g(x)$. Define $h(x, y) = \frac{1}{2}\sum_{i=0}^{3} B(g^{-i}x, y)\varsigma^i$. Then $h(x, y)$ is a totally positive definite hermitian form with respect to complex conjugation.

Proposition 4.5. $h(x, y) = B(x, y) - B(gx, y)\varsigma$.

Proof.

$$
\begin{aligned}
h(x, y) &= \frac{1}{2}\{B(x, y) + B(g^{-1}x, y)\varsigma + B(-x, y)(-1) + B(gx, y)(-\varsigma)\} \\
&= \frac{1}{2}\{2B(x, y) + B((g^{-1} - g)x, y)\varsigma\} \\
&= B(x, y) + \frac{1}{2}B(g(g^{-2} - 1)x, y)\varsigma \\
&= B(x, y) - B(gx, y)\varsigma.
\end{aligned}
$$

□

By Proposition 4.5, $h(x, y) \in \mathbb{Z}[\varsigma]$ for any x and $y \in M$. Thus M has a hermitian $\mathbb{Z}[\varsigma]$ lattice structure and W has a hermitian vector space structure. Let us denote them by $\mathcal{M}$ and $\mathcal{W}$ respectively.

Proposition 4.6. Let $y \in \mathcal{W}$. Then $h(x, y) \in \mathbb{Z}[\varsigma]$ for every $x \in \mathcal{M}$ if any only if $y \in \mathcal{M}$.

Proof. First assume that $h(x, y) \in \mathbb{Z}[\varsigma]$ for every $x \in \mathcal{M} = M$. Then by Proposition 4.5, $B(x, y) \in \mathbb{Z}$ for every $x \in \mathcal{M} = M$. Since M is unimodular with respect to B, we have $y \in M = \mathcal{M}$. Conversely, it is clear that $h(x, y) \in \mathbb{Z}[\varsigma]$ for any x and $y \in \mathcal{M}$.   □

Proposition 4.7. Let M and N be lattices in G_L of type $IR(4)$. Let $\mathcal{M}$ and $\mathcal{N}$ be the hermitian $\mathbb{Z}[\varsigma]$-lattices constructed from M and N, h_M and h_N be the hermitian form of $\mathcal{M}$ and $\mathcal{N}$, respectively. Suppose $(\mathcal{M}, h_M)$ is isometric to $(\mathcal{N}, h_N)$ as a hermitian $\mathbb{Z}[\varsigma]$-lattices. Then (M, B) is isometric to (N, B) as quadratic lattices.

Proof. The proof is similar to that of Proposition 4.2.   □

Corollary 4.8.

$$|U(\mathcal{M})| \le |O(M)|$$

Proof. In Proposition 4.7 take $N = M$.   □

Lemma 4.9. There is exactly one genus G_M of totally positive definite unimodular hermitian $\mathbb{Z}[\sqrt{-1}]$ lattice with norm $n(M) = 2\mathbb{Z}[\sqrt{-1}]$ and of rank $\frac{m}{2}$. Moreover if L is an even unimodular lattice then we have $\omega_{IR(4)} \leq \omega(M)$.

Proof. Proposition 3.2 in [14], Propositions 10.3, 9.2, 10.4 and Theorem 7.1 in [7] show that there is exactly one genus of such lattices. By Proposition 4.6 and arguments similar to those in [8], §82 F and §82 G, we can show that every lattice $\mathcal{N}$ constructed from a lattice N in the genus G_L of type $IR(4)$ is unimodular. Since $B(g_N x, x) = B(g_N^2 x, g_N x) = B(-x, g_N x)$, we have $h(x, x) = B(x, x)$ for any x in N. Therefore if L is even unimodular, then by Proposition 4.5, $n(\mathcal{N}) \subset 2\mathbb{Z}[\sqrt{-1}]$. Therefore $\mathcal{N} \in G_M$. By Corollary 4.8 we have $\omega_{IR(4)} \leq \omega(M)$. $\qquad\qquad\square$

Lemma 4.10. There are exactly two genera G_M and G_N of totally positive definite unimodular hermitian $\mathbb{Z}[\sqrt{-1}]$ - lattice of rank $r = \frac{m}{2}$ with norm $\mathbb{Z}[\sqrt{-1}]$. Moreover if L is odd unimodular lattice then we have

$$\omega_{IR(4)} \leq \omega(M) + \omega(N)$$

Proof. The same type of argument as in the proof of Lemma 4.9 gives Lemma 4.10. $\qquad\square$

CHAPTER III
LOCAL DENSITIES

§5. Local densities of hermitian lattices.

In this section we use the notation given in §2. First we introduce some more definitions and notation (see also §2).

Let M be a hermitian S_p-lattice on W. Define the dual lattice $M\#$ of M to be the set $\{y \in W \mid h(y, M) \subseteq S_p\}$ (clearly this set is a lattice on W). If P does not split in E then define $ord_p(A)$ to be the order of the generator of A with respect to P for any fractional ideal in E_p. If P splits in E then $ord_p(A \otimes R_p)$ is defined to be the order of the generator of $A_{\mathbb{P}}$ with respect to P for any fractional ideal A in E, where $\mathbb{P}$ is a prime ideal in S such that $\mathbb{P} \mid P$. Let Δ_p be the discriminant of S_p over R_p. Let N be a hermitian S_p-lattice. Let r be the rank of M and s be the rank of N, respectively, where $r \geq s$. Define $A_{p^\mu}(M, N)$ to be the number of S_p-linear maps $\sigma : N \to M$ which are distinct modulo $P^\mu M$ and satisfy $h_M(\sigma x, \sigma y) \equiv h_N(x, y) \, mod \, P^\mu S_p$ for every x and y in N. Then we have the next two propositions by Rehmann.

Proposition 5.1 (Hilfssatz 5.3 [12])

(i) If $P \nmid 2$ or P splits in E then

$$A_{p^{\mu+1}}(M, N) = p^{fps(2r-s)} A_{p^\mu}(M, N)$$

for any $\mu \geq 2\alpha + 1$ with $\alpha \geq -ord_p(sN\#)$, where p is a rational prime number such that $P \mid p$.

(ii) If $P \mid 2$ and P does not split in E then $A_{p^{\mu+1}}(M, N) = p^{fps(2r-s)} A_{p^\mu}(M, N)$ for any $\mu \geq 2\alpha + 1$ with $\alpha \geq ord_p(\sqrt{\Delta_p}(sN\#)^{-1})$.

Proposition 5.2 (Hilfssatz 6.1 [12]). Let P be an unramified prime ideal in K. Then the local density $\beta_P(M)$ of a unimodular hermitian S_p-lattice M of rank r is given by the following:

$$\beta_P(M) = \prod_{i=1}^{r} (1 - p^{-fpi}) \text{ for } P \text{ which splits in } K.$$

$$\beta_P(M) = \prod_{i=1}^{r} (1 - (-1)^i p^{-fpi}) \text{ for } P \text{ which remains prime in } K.$$

22

Rehmann also gave some estimations for the upper bounds of the local densities of unimodular lattices and λS_P-modular lattices at the ramified primes (See Hilfssatz 6.2, 6.3 in [12]). It seems that there is some error in Hilfsatz 6.3 [12]. Hilfssatz 6.3 shows that the local density of a λS_P-modular hermitian S_P-lattice at the prime P above q is bounded by $q^{\frac{1}{2}r(r-1)}$ from above. This apparently contradicts our Proposition 5.8.

In the following we will give local densities of some hermitian S_P-lattices at the ramified prime ideal P in K. (That is $P \mid q$ if q is odd prime and $P \mid 2$ if $q = 4$.) We assume scales of the lattices in this section are ideals in S_P.

Proposition 5.3. (Analog of 82:15 in [8].) Let M be a lattice on a hermitian space $\mathcal{W}$ over E_P. Let J be a modular sublattice of M such that $h(M, J) \subseteq sJ$. Then there exists a sublattice N of M such that $M = J \perp N$.

Proof. Let $\mathcal{U}$ be the orthogonal compliment of $E_P J$ in $\mathcal{W}$. Let $x \in M$ then $x = y + z$ with some $y \in E_P J$ and $z \in \mathcal{U}$. Then $h(y, J) = h(x - z, J) = h(x, J) \subseteq h(M, J) \subseteq sJ$. Since J is a modular lattice, $y \in J \subset M$ and therefore $z = x - y \in M \cap U$. Take $N = M \cap U$. $\qquad\square$

Proposition 5.4. Let r and ℓ be integers $r \geq \ell \geq 1$. Let M and N are hermitian lattice over S_P of rank r and ℓ respectively. Let H_M and H_N be the hermitian matrices over S_P with respect to certain bases of M and N respectively. Assume that there exists an integer μ_0 such that for any hermitian (ℓ, ℓ)-matrix C and any integer $\mu \geq \mu_0$, $C \equiv H_N$ modulo $P^\mu S_P$ implies that the lattice defined by C is isometric to N as S_P-lattices. Then $q^{2\ell r} A_{P^\mu}(M, N) = q^{\ell^2} A_{P^{\mu+1}}(M, N)$ for any $\mu \geq \mu_0$.

Proof. Let $X_j, j = 1, \cdots, A_{P^\mu}(M, N)$ be all the mod (r, ℓ)-matrices which are distinct modulo $P^\mu S_P$ satisfying ${}^t\overline{X}_j H_M X_j \equiv H_N (mod\ P^\mu S_P)$. For each j let $Y_i^j, i = 1, \cdots, q^{2\ell r}$ be all (r, ℓ)-matrices which are distinct modulo $P^{\mu+1} S_P$ satisfying $Y_i^j \equiv X_j$ $(mod\ P^\mu S_P)$. Then ${}^t\overline{Y}_i^j H_M Y_i^j \equiv H_N (mod\ P^\mu S_P)$. Let $B_t, t = 1, \cdots q^{\ell^2}$ be all the (ℓ, ℓ)-hermitian matrices over S_P which are distinct modulo $P^{\mu+1} S_P$ satisfying $B_t \equiv H_N (mod\ P^\mu S_P)$. Then ${}^t\overline{Y}_i^j H_M Y_i^j \equiv B_t (mod\ P^{\mu+1} S_P)$ with some t. Therefore we have

$$q^{2\ell r} A_{P^\mu}(M, N) \leq \sum_{i=1}^{q^{\ell^2}} A_{P^{\mu+1}}(M, B_i)$$ (here B_i denote the S_P-lattice defined

by the hermitian matrix B_i). On the other hand, any matrix such that ${}^t\overline{Y} H_M Y \equiv B_i$ $(mod\ P^{\mu+1} S_P)$ gives ${}^t\overline{Y} H_M Y \equiv H_N (mod\ P^\mu S_P)$. Therefore $Y \equiv Y_i^j (mod\ P^{\mu+1} S_P)$ with some Y_i^j. This shows that $q^{2\ell r} A_{P^\mu}(M, N) = \sum_{i=1}^{q^{\ell^2}} A_{P^{\mu+1}}(M, B_i)$. By assumption B_i is isometric to N therefore $A_{P^{\mu+1}}(M, B_i) = A_{P^{\mu+1}}(M, N)$ for $i = 1, \cdots, q^{\ell^2}$. This completes the proof. $\qquad\square$

Case $q =$ odd prime and $P \mid q$

Proposition 5.5. Let $M = J \perp N$ with some $\lambda^s S_p$-modular lattice J ($s = 0$ or 1) such that $sM = sJ$. Then $A_{P\mu}(M, M) = q^{st(r-t)} A_{P\mu}(M, J) A_{P\mu}(N, N)$, where $\mathrm{rank}_{S_p} M = r$, $\mathrm{rank}_{S_p} J = t$ and μ is any integer ≥ 1.

Proof. Let H_M, H_J and H_N be the hermitian matrices of M, J and N corresponding to certain basis of M respectively. Therefore $H_M = \begin{pmatrix} H_J & 0 \\ 0 & H_N \end{pmatrix}$. Let $C = (c_{ij})$ be a (r, t)-matrix over S_p such that ${}^t\overline{C} H_M C \equiv H_J (\mathrm{mod}\ P^\mu S_p)$. Then ${}^t\overline{C} H_M C$ corresponds to a sublattice J_1 of M. Since ${}^t\overline{C} H_M C \equiv H_J(\mathrm{mod}\ P^\mu S_p)$, J_1 is isometric to J by Theorem 8.2 in [7]. Therefore by Proposition 5.3 J_1 splits M. That is there exists a (r, r)-matrix $A = (a_{ij})$ such that $a_{ij} = c_{ij}$ for $i = 1, \cdots r, j = 1, \cdots, t$ and ${}^t\overline{A} H_M A \equiv H_M\ (\mathrm{mod}\ P^\mu S_p)$. Let $X = (x_{ij})$ be a (r, r)-matrix such that $x_{ij} \equiv c_{ij}(\mathrm{mod}\ P^\mu S_p), i = 1, \cdots, r, j = 1, \cdots, t$, and ${}^t\overline{X} H_M X \equiv H_M(\mathrm{mod}\ P^\mu S_p)$. Then

$$A^{-1}X \equiv \begin{pmatrix} 1 & & & 0 \\ & \ddots & & B \\ 0 & & 1 & \\ & 0 & & X_1 \end{pmatrix} (\mathrm{mod}\ P^\mu S_p) \text{ with some } (t, n)\text{-matrix } B \text{ and } (n, n)\text{-}$$

matrix X_1, where $n = r - t$. Then we have ${}^t\overline{X} H_M X \equiv \begin{pmatrix} H_J & H_J B \\ {}^t\overline{B} H_J & {}^t\overline{X}_1 H_N X_1 \end{pmatrix} \equiv$

$\begin{pmatrix} H_J & 0 \\ 0 & H_N \end{pmatrix}$ $(\mathrm{mod}\ P^\mu S_p)$. Hence we have $H_J B \equiv 0(\mathrm{mod}\ P^\mu S_p)$. By Proposition 8.1

in [7] we may assume $H_J = \begin{pmatrix} 1 & & \\ & \ddots & \\ & & 1 \\ & & & dJ \end{pmatrix}$ if $s = 0(dJ = $ a unit in $K_p)$ and

$$H_J = \begin{pmatrix} 0 & \lambda & & & 0 \\ -\lambda & 0 & & & \\ & & \ddots & & \\ 0 & & & 0 & \lambda \\ & & & -\lambda & 0 \end{pmatrix} \text{ if } s = 1. \text{ Therefore } B \equiv 0\ (\mathrm{mod}\ P^\mu S_p) \text{ if } s = 0,$$

and $B \equiv 0(\mathrm{mod}\ \lambda^{2\mu-1} S_p)$ if $s = 1$. Therefore the number of choices for B is $q^{st(r-t)}$. This completes the proof. $\qquad \square$

Proposition 5.6. Let $M = \langle \epsilon \rangle \perp \langle 1 \rangle \perp \cdots \perp \langle 1 \rangle$ of $\mathrm{rank}_{S_p} M = r$. Then we have

(i) if $r = $ even then $A_P(M, \langle \epsilon \rangle) = q^{2r-1}(1 - (\frac{(-1)^{\frac{r}{2}}\epsilon}{p})q^{-\frac{r}{2}})$

(ii) if $r = $ odd then $A_P(M, \langle \epsilon \rangle) = q^{2r-1}(1 + (\frac{(-1)^{\frac{r-1}{2}}}{p})q^{-\frac{r-1}{2}})$.

Proof. The number $A_P(M, \langle \epsilon \rangle)$ is the same as the number of vectors $\mathbf{a} = {}^t(a_1, \cdots a_r)$ which are distinct modulo $P S_p$ satisfying

$$(3) \qquad\qquad\qquad {}^t\overline{\mathbf{a}} H \mathbf{a} \equiv \epsilon(\mathrm{mod}\ P S_p).$$

Since S_p is generated by 1 and λ over R_p, $a_i = \xi_i + \lambda \eta_i$ with some $\xi_i, \eta_i \in R_p$. Then the equation (3) is equivalent to

$$(4) \qquad\qquad\qquad \epsilon \xi_i^2 + \sum_{i=2}^{r} \xi_i^2 \equiv \epsilon\ (\mathrm{mod}\ P R_p)$$

with η_i chosen arbitrarily $mod\ PS_P$. The number of the solutions of (4) is well-known (see Hilfssatz 12 in [15]) and is $q^{r-1}(1 - (\frac{(-1)^{\frac{r}{2}}\epsilon}{p})q^{-\frac{r}{2}})$ if r = even, and $q^{r-1}(1 + (\frac{(-1)^{\frac{r-1}{2}}}{p})q^{-\frac{r-1}{2}})$ if r = odd. This gives the Proposition 5.6. $\qquad\square$

Proposition 5.7. Let M be a unimodular hermitian S_P-lattice, $rank_{S_P} M = r$ and $dM = \epsilon$. Then the local density of M is given by the following:

(i) if r = even, then

$$\beta_P(M) = 2(1 - \frac{(-1)^{\frac{r}{2}}\epsilon}{p})q^{-\frac{r}{2}})\prod_{i=1}^{\frac{r-2}{2}}(1 - q^{-2i}),$$

if r = odd, then

$$\beta_P(M) = 2\prod_{i=1}^{\frac{r-1}{2}}(1 - q^{-2i}).$$

Proof. By Proposition 8.1 in [7], $M \cong \langle\epsilon\rangle \perp \langle 1\rangle \perp \cdots \perp \langle 1\rangle$. By Proposition 5.1, $\beta_P(M) = q^{-r^2}A_P(M, M)$. By Proposition 5.5, $A_P(M, M) = A_P(M, \langle\epsilon\rangle)A_P(N, N)$, where $N = \langle 1\rangle \perp \cdots \perp \langle 1\rangle$ and $rank_{S_P} N = r - 1$. Therefore by Proposition 5.6, if r = even then

$$A_P(M, M) = q^{2r-1}(1 - \frac{(-1)^{\frac{r}{2}}\epsilon}{p})q^{-\frac{r}{2}})A_P(N, N),$$

and if r = odd then

$$A_P(M, M) = q^{2r-1}(1 + \frac{(-1)^{\frac{r-1}{2}}}{p})q^{-\frac{r-1}{2}})A_P(N, N).$$

Inductive computations give the Proposition 5.7. $\qquad\square$

Proposition 5.8 Let M be a λS_P-modular hermitian lattice of $rank_{S_P} M = r$. Then the local density at P is given by

$$\beta_P(M) = q^{\frac{1}{2}r(r+1)}\prod_{i=1}^{\frac{r}{2}}(1 - q^{-2i})$$

Proof. By Proposition 8.1 in [7], r has to be even and M is isometric to $\begin{pmatrix} & \lambda \\ -\lambda & \end{pmatrix} \perp \begin{pmatrix} & \lambda \\ -\lambda & \end{pmatrix} \perp \cdots \perp \begin{pmatrix} & \lambda \\ -\lambda & \end{pmatrix}$. Let H_r be the (r, r)-matrix

$$\begin{pmatrix} 0 & \lambda & & & & \\ -\lambda & 0 & & & & \\ & & 0 & \lambda & & \\ & & -\lambda & 0 & & 0 \\ & & & & \ddots & \\ & 0 & & & & 0 & \lambda \\ & & & & & -\lambda & 0 \end{pmatrix}.$$

Let C be a $(r, 2)$-matrix ${}^t\begin{pmatrix} a_{11}b_{11} & \cdots & a_{\frac{r}{2},1}b_{\frac{r}{2},1} \\ a_{12}b_{12} & \cdots & a_{\frac{r}{2},2}b_{\frac{r}{2},2} \end{pmatrix}$ such that

(5) $${}^t\overline{C}H_r C \equiv H_2 \ (mod\ PS_P).$$

Then number of the solutions of (5) which are distinct modulo PS_P is the same as the number of the solutions which satisfy the following three equations (6), (7) and (8).

$$(6) \qquad \lambda \sum_{i=1}^{\frac{r}{2}} (\bar{a}_{i1} b_{i1} - a_{i1} \bar{b}_{i1}) \equiv 0 (mod\ PS_P).$$

$$(7) \qquad \lambda \sum_{i=1}^{\frac{r}{2}} (\bar{a}_{i2} b_{i2} - a_{i2} \bar{b}_{i2}) \equiv 0 (mod\ PS_P).$$

$$(8) \qquad \lambda \sum_{i=1}^{\frac{r}{2}} (\bar{a}_{i1} b_{i2} - \bar{b}_{i1} a_{i2}) \equiv \lambda (mod\ PS_P).$$

Let $a_{ij} = a_{ij}^{(0)} + a_{ij}^{(1)} \lambda, b_{ij} = b_{ij}^{(0)} + b_{ij}^{(1)} \lambda$ with $a_{ij}^{(\ell)}, b_{ij}^{(\ell)} \in R_P$. Then the number of solutions which satisfy (6), (7) and (8) is equal to the number of solutions which satisfy the following three equations.

$$(9) \qquad \lambda^2 \sum_{i=1}^{\frac{r}{2}} (a_{i1}^{(0)} b_{i1}^{(1)} - a_{i1}^{(1)} b_{i1}^{(0)}) \equiv 0\ (mod\ PR_P),$$

$$(10) \qquad \lambda^2 \sum_{i=1}^{\frac{r}{2}} (a_{i2}^{(0)} b_{i2}^{(1)} - a_{i2}^{(1)} b_{i2}^{(0)}) \equiv 0\ (mod\ PR_P),$$

$$(11) \qquad \lambda \sum_{i=1}^{\frac{r}{2}} (a_{i1}^{(0)} b_{i2}^{(0)} - b_{i1}^{(0)} a_{i2}^{(0)}) + \lambda^3 \sum_{\lambda=1}^{\frac{r}{2}} (b_{i1}^{(1)} a_{i2}^{(1)} - a_{i1}^{(1)} b_{i2}^{(1)})$$

$$+ \lambda^2 \sum_{i=1}^{\frac{r}{2}} (a_{i1}^{(0)} b_{i2}^{(1)} - a_{i1}^{(1)} b_{i2}^{(0)} - b_{i1}^{(0)} a_{i2}^{(1)} + b_{i1}^{(1)} a_{i2}^{(0)}) \equiv \lambda\ (mod\ PR_P)$$

Since $\lambda^2 R_P = P$, equations (9) and (10) are satisfied automatically. Equation (11) is equivalent to

$$(12) \qquad \sum_{i=1}^{\frac{r}{2}} (a_{i1}^{(0)} b_{i2}^{(0)} - b_{i1}^{(0)} a_{i2}^{(0)}) \equiv 1\ (mod\ PR_P).$$

Then by (12) $a_{ik}, i = 1, \cdots, \frac{r}{2}, k = 1, 2$, are not in PR_P at the same time. For any such $\{a_{ik}^{(0)}\} i = 1, \cdots, \frac{r}{2}, k = 1, 2$ fixed, the number of the solution of (12) is q^{r-1} . The number of such $\{a_{ik}^{(0)}\}_{i=1,\cdots,\frac{r}{2}, k=1,2}$ is $q^r - 1$. Therefore we have $A_P(M, H_2) = q^{2r} q^{r-1} (q^r - 1) = q^{4r-1}(1 - q^{-r})$. By Proposition 5.5 $A_P(M, M) = q^{2(r-2)} A_P(M, H_2)$
$A_P(H_{r-2}, H_{r-2}) = q^{6r-5}(1 - q^{-r}) A_P(H_{r-2}, H_{r-2})$, where H_ℓ also denotes the lattice defined by the hermitian matrix H_ℓ . By inductive computation we have $A_P(M, M) = q^{\frac{3}{2}r^2 + \frac{1}{2}r} \prod_{i=1}^{\frac{r}{2}} (1 - q^{-2i})$. Therefore $\beta_P(M) = q^{-r^2} A_P(M, M) = q^{\frac{1}{2}r(r+1)} \prod_{i=1}^{\frac{r}{2}} (1 - q^{-2i})$. $\square$

Proposition 5.9. Let $M \cong \langle \epsilon \rangle \perp \langle 1 \rangle \perp \cdots \perp \langle 1 \rangle \perp H_{r-\rho}$, where $rank_{S_\rho} M = r$, $r - \rho$ is even and $0 \leq \rho \leq r$. Then the local density of M is given by the following equations.

(i) If ρ = even, then

$$\beta_p(M) = \frac{2q^{\frac{1}{2}(r-\rho)(r-\rho+1)} \prod_{i=1}^{\frac{\rho}{2}}(1-q^{-2i}) \prod_{i=1}^{\frac{r-\rho}{2}}(1-q^{-2i})}{1+(\frac{(-1)^{\frac{\rho}{2}}\epsilon}{q})q^{-\frac{\rho}{2}}}$$

(ii) If ρ = odd, then

$$\beta_p(M) = 2q^{\frac{1}{2}(r-\rho)(r-\rho+1)} \prod_{i=1}^{\frac{\rho-1}{2}}(1-q^{-2i}) \prod_{i=1}^{\frac{r-\rho}{2}}(1-q^{-2i}).$$

Proof. If $\rho = 0$ then it is already shown in Proposition 5.8. Therefore we may assume $\rho \geq 1$. Let $\mathbf{a} = {}^t(a_1, \cdots, a_r)$ be a vector over S_p such that

(13)

$${}^t\mathbf{a} \begin{pmatrix} \epsilon & & & & & & & \\ & 1 & & & & & 0 & \\ & & \ddots & & & & & \\ & & & 1 & & & & \\ & & & & 0 & \lambda & & \\ & & & & -\lambda & 0 & & \\ & 0 & & & & & \ddots & \\ & & & & & & & 0 & \lambda \\ & & & & & & & -\lambda & 0 \end{pmatrix} \mathbf{a} \equiv \epsilon \quad (mod\, p\, S_p).$$

Let $a_i = a_{i0} + \lambda_{i1}$ with $a_{i0}, a_{i1}, \in R_p$ for $i = 1, \cdots, r$. Then (13) is equivalent to

(14)
$$\epsilon a_{10}^2 + \sum_{i=2}^{\rho} a_{i0}^2 \equiv \epsilon(mod\, P\, R_p).$$

The number of the solutions of (14) is well known (see Hilfssatz 12 in [15]) and is $q^{\rho-1}(1-(\frac{(-1)^{\frac{\rho}{2}}\epsilon}{p})q^{-\frac{\rho}{2}})$ if ρ = even, and $q^{\rho-1}(1+(\frac{(-1)^{\frac{\rho-1}{2}}}{p})q^{-\frac{\rho-1}{2}})$ if ρ = odd.

Therefore $A_p(M, \langle\epsilon\rangle) = q^{2r-1}(1-(\frac{(-1)^{\frac{\rho}{2}}\epsilon}{p})q^{-\frac{\rho}{2}})$ if ρ = even, and $A_p(M, \langle\epsilon\rangle) = q^{2r-1}(1+(\frac{(-1)^{\frac{\rho-1}{2}}}{p})q^{-\frac{\rho-1}{2}})$ if ρ = odd. On the other hand, if $a \equiv \epsilon(P)$ and $a \in R_p$ then $\langle a \rangle \cong \langle\epsilon\rangle$ as hermitian S_p-lattices. Therefore by Proposition 5.4 we have $q^{2r}A_{p^\mu}(M, \langle\epsilon\rangle) = qA_{p^{\mu+1}}(M, \langle\epsilon\rangle)$ for $\mu \geq 1$. Therefore $A_{p^3}(M, \langle\epsilon\rangle) = q^{4r-2}A_p(M, \langle\epsilon\rangle)$.

Hence, we have by Proposition 5.5

$$A_{p^3}(M, M) = A_{p^3}(M, \langle\epsilon\rangle)A_{p^3}(N, N)$$
$$= q^{6r-3}(1-(\frac{(-1)^{\frac{\rho}{2}}\epsilon}{q})q^{-\frac{\rho}{2}})A_{p^3}(N, N) \text{ if } \rho \text{ even}$$

and

$$= q^{6r-3}(1+(\frac{(-1)^{\frac{\rho-1}{2}}}{q})q^{-\frac{\rho-1}{2}})A_{p^3}(N, N) \text{ if } \rho \text{ odd,}$$

where $N = \langle 1 \rangle \perp \cdots \perp \langle 1 \rangle \perp H_{r-\rho}$ and rank $_{S_\rho} N = r - 1$. Then by inductive computation and Proposition 5.8 we can complete the proof.   □

Case $q = 4$ (dyadic ramified case) and $P \mid 2$.

In this case $E = \mathbb{Q}(\sqrt{-1}), K = \mathbb{Q}, K_2 = \mathbb{Q}_2, R_2 = \mathbb{Z}_2, E_2 = \mathbb{Q}_2(\sqrt{-1})$ and $S_2 = \mathbb{Z}_2[\sqrt{-1}]$.

Let H_i be the uniodular hermitian S_2-lattice $\begin{pmatrix} 0 & 1 \\ 1 & 0 \end{pmatrix} \perp \cdots \perp \begin{pmatrix} 0 & 1 \\ 1 & 0 \end{pmatrix}$ where

rank $_{S_2} H_i = i$. We also denote the (i, i)-hermitian matrix $\begin{pmatrix} 0 & 1 & & & \\ 1 & 0 & & & 0 \\ & & \ddots & & \\ 0 & & & & 1 \\ & & & 1 & 0 \end{pmatrix}$ by

H_i.

Proposition 5.10. Let C be a hermitian (r, r)-matrix over S_2 such that $C \equiv H_r$ $(mod\ 2S_2)$, then the lattice defined by C is isometric to H_i.

Proof. Let M be the lattice defined by C. Clearly M is unimodular and $n(M) \subseteq 2S_2$. By Proposition 10.3 in [7], $M = J \perp H_{r-2}$ with some unimodular lattice J of rank 2 and $n(J) = n(M)$. By Proposition 9.1 in [7], $n(J) \supseteq n(H_2) = 2S_2$. Therefore by Proposition 9.2 in [7], $J \cong H_2$.   □

Proposition 5.11. Let $M = H_2 \perp N$, where N is a unimodular hermitian S_2-lattice of rank $r - 2$. Then we have $A_{2\mu}(M, M) = A_{2\mu}(M, H_2)A_{2\mu}(N, N)$ for any $\mu \geq 1$.

Proof. Let H_M and H_N be the hermitian matrices of the lattices M and N with respect to certain base of M, respectively. Then we may assume $H_M = \begin{pmatrix} H_2 & 0 \\ 0 & H_N \end{pmatrix}$. Let C be a $(r, 2)$-matrix such that ${}^t\overline{C}H_M C \equiv H_2\ (mod\ 2^\mu S_2)$. Then ${}^t\overline{C}H_M C$ defines a sublattice J of M. By Proposition 5.10 $J \cong H_2$. By Proposition 5.3, J splits M, i.e., $M = J \perp K$ with some unimodular sublattice K of M. By Proposition 9.3 in [7], $N \cong K$. Therefore there exists a (r, r)-matrix $A = (a_{ij})$ such that $a_{ij} = c_{ij}, i = 1, \cdots, r, j = 1, 2$, and ${}^t\overline{A}H_M A \equiv H_M(mod\ 2^\mu S_2)$.

Let $X = (x_{ij})$ be a (r, r)-matrix over S_2 such that $x_{ij} \equiv c_{ij}(mod\ 2^\mu S_2), i = 1, \cdots, r, j = 1, 2$, and ${}^t\overline{X}H_M X = H_M(mod\ 2^\mu S_2)$. Then $A^{-1}X \equiv \begin{pmatrix} H_2 & B \\ 0 & X_1 \end{pmatrix}$ $(mod\ 2^\mu S_2)$ with some $(2, r-2)$-matrix B over S_2. We have ${}^t\overline{X}H_M X \equiv \begin{pmatrix} H_2 & H_2 B \\ {}^t\overline{B}H_2 & {}^t\overline{X_1}H_N X_1 \end{pmatrix}$. Therefore we have $B \equiv 0(mod\ 2^\mu S_2)$ and we have the Proposition.   □

Proposition 5.12. Let $M \cong H_r$. Then

$$\beta_2(M) = 2^r \prod_{i=1}^{\frac{r}{2}} (1 - 2^{-2i}).$$

Proof. By Propositions 5.10 and 5.4 we have $\beta_2(M) = 2^{-r^2} A_2(H_r, H_r)$. By Proposition 5.11 we have $A_2(H_r, H_r) = A_2(H_r, H_2) A_2(H_{r-2}, H_{r-2})$. Let

$$C = \begin{pmatrix} a_{11}b_{11} & \cdots & a_{\frac{r}{2},1}b_{\frac{r}{2},1} \\ a_{12}b_{12} & \cdots & a_{\frac{r}{2},2}b_{\frac{r}{2},2} \end{pmatrix}$$

be a $(r, 2)$ -matrix over S_2 such that

$$(15) \qquad\qquad {}^t\overline{C}H_r C \equiv H_2 \ (mod \, 2S_2).$$

Then (15) is equivalent to the following equations:

$$(16) \qquad\qquad \sum_{i=1}^{\frac{r}{2}} (a_{i1}\overline{b}_{i1} + \overline{a}_{i1}b_{i1}) \equiv 0 \ (mod \, 2S_2)$$

$$(17) \qquad\qquad \sum_{i=1}^{\frac{r}{2}} (a_{i2}\overline{b}_{i2} + \overline{a}_{i2}b_{i2}) \equiv 0 \ (mod \, 2S_2)$$

$$(18) \qquad\qquad \sum_{i=1}^{\frac{r}{2}} (a_{i2}\overline{b}_{i1} + \overline{a}_{i1}b_{i2}) \equiv 1 \ (mod \, 2S_2)$$

Let

$$a_{ij} = a_{ij}^{(0)} + a_{ij}^{(1)}\sqrt{-1}$$
$$b_{ij} = b_{ij}^{(0)} + b_{ij}^{(1)}\sqrt{-1}$$

where $a_{ij}^{(k)}, b_{ij}^{(k)} \in \mathbb{Z}_2$ $i = 1, \cdots, \frac{r}{2}; \ j = 1, 2; k = 0, 1$.

It is easily seen that (16) and (17) are satisfied for any $a_{ij}^{(k)}, b_{ij}^{(k)}$ and that (18) is equivalent to

$$(19) \qquad \sum_{i=1}^{\frac{r}{2}} (a_{i2}^{(0)}b_{i1}^{(0)} + a_{i1}^{(0)}b_{i2}^{(0)} + a_{i2}^{(1)}b_{i1}^{(1)} + a_{i1}^{(1)}b_{i2}^{(1)})$$
$$+ \sqrt{-1} \sum_{i=1}^{\frac{r}{2}} a_{i2}^{(1)}b_{i1}^{(0)} - a_{i2}^{(0)}b_{i1}^{(1)} - a_{i1}^{(1)}b_{i2}^{(0)} + a_{i1}^{(0)}b_{i2}^{(1)})$$
$$\equiv 1 \ (mod \, 2\mathbb{Z}_2)$$

and hence (19) is equivalent to the following two equations:

$$(20) \qquad \sum_{i=1}^{\frac{r}{2}} (a_{i2}^{(0)}b_{i1}^{(0)} + a_{i1}^{(0)}b_{i2}^{(0)} + a_{i2}^{(1)}b_{i1}^{(1)} + a_{i1}^{(1)}b_{i2}^{(1)}) \equiv 1 \ (mod \, 2\mathbb{Z}_2)$$

$$(21) \qquad \sum_{i=1}^{\frac{r}{2}} (a_{i2}^{(1)}b_{i1}^{(0)} - a_{i1}^{(1)}b_{i2}^{(0)} - a_{i2}^{(0)}b_{i1}^{(1)} + a_{i1}^{(0)}b_{i2}^{(1)}) \equiv 0 \ (mod \, 2\mathbb{Z}_2)$$

Let $A = \begin{pmatrix} \cdots & a_{i2}^{(0)}, & a_{i1}^{(0)}, & a_{i2}^{(1)}, & a_{i1}^{(1)}, & \cdots \\ \cdots & a_{i2}^{(1)}, & a_{i1}^{(1)}, & a_{i2}^{(0)}, & a_{i1}^{(0)}, & \cdots \end{pmatrix}$ and $\mathbf{b} =^t (\cdots b_{i1}^{(0)} b_{i2}^{(0)} b_{i1}^{(1)} b_{i2}^{(1)} \cdots)$

Then (20) and (21) are equivalent to

$$(22) \qquad\qquad Ab \equiv \begin{pmatrix} 1 \\ 0 \end{pmatrix} \ (mod\ 2\mathbf{Z}_2).$$

Therefore, the rank of A must be 2. Hence, for a fixed such A, the number of solutions of $mod\ 2\mathbf{Z}_2$ is 2^{2r-2}. It is easy to see that the number of such matrices A of rank 2 is $2^{2r} - 2^r$. Therefore we have $A_2(H_r, H_2) = 2^{4r-2}(1 - 2^{-r})$ and $A_2(H_r, H_r) = 2^{4r-2}(1 - 2^{-r})A_2(H_{r-2}, H_{r-2})$. By inductive computation we have

$$A_2(H_r, H_r) = 2^{r^2+r} \prod_{i=1}^{\frac{r}{2}} (1 - 2^{-2i}).$$

This completes the proof.   □

Proposition 5.13. Let $\mathcal{M} = \langle \epsilon \rangle \perp H_{r-1}$ with some odd integer $r \geq 1$ and a unit ϵ in $\mathbf{Z}_2$. Then we have

$$\beta_2(\mathcal{M}) = 2 \prod_{i=1}^{\frac{r-1}{2}} (1 - 2^{-2i}).$$

Proof. If $r = 1$, then by Propositon 5.1

$$\beta_2(\langle \epsilon \rangle) = 2^{-3} A_{2^3}(\langle \epsilon \rangle, \langle \epsilon \rangle).$$

It is easy to see $A_{2^3}(\langle \epsilon \rangle, \langle \epsilon \rangle) = 2^4$, i.e., $\beta_2(\langle \epsilon \rangle) = 2$. Therefore we may asume $r \geq 3$. Let $\mathcal{N} = \langle \epsilon \rangle \perp H_{r-3}$. Then Proposition 5.11 gives

$$A_{2^\mu}(\mathcal{M}, \mathcal{M}) = A_{2^\mu}(\mathcal{M}, H_2)A_{2^\mu}(\mathcal{N}, \mathcal{N})$$

for any $\mu \geq 1$. First we evaluate $A_2(\mathcal{M}, H_2)$. Let
$C = \begin{pmatrix} a_1 a_{11} b_{11} & \cdots & a_{\frac{r-1}{2},1} b_{\frac{r-1}{2},1} \\ a_2 a_{12} b_{12} & \cdots & a_{\frac{r-1}{2},2} b_{\frac{r-1}{2},2} \end{pmatrix}$ be a $(r, 2)$-matrix over S_2 such that

$$(23) \qquad\qquad {}^t\overline{C}HC \equiv H_2 \ (mod\ 2S_2).$$

Then (23) is equivalent to the following three equations:

$$(24) \qquad\qquad \epsilon a_1 \overline{a}_1 + \sum_{i=1}^{\frac{r-1}{2}} (a_{i1} \overline{b}_{i1} + \overline{a}_{i1} b_{i1}) \equiv 0 \ (mod\ 2S_2)$$

$$(25) \qquad\qquad \epsilon a_2 \overline{a}_2 + \sum_{i=2}^{\frac{r-1}{2}} (a_{i2} \overline{b}_{i2} + \overline{a}_{i2} b_{i2}) \equiv 0 \ (mod\ 2S_2)$$

$$(26) \qquad\qquad \epsilon \overline{a}_1 a_2 + \sum_{i=1}^{\frac{r-1}{2}} (a_{i2} \overline{b}_{i1} + \overline{a}_{i1} b_{i2}) \equiv 1 \ (mod\ 2S_2)$$

Put

$$a_i = a_i^{(0)} + a_i^{(1)}\sqrt{-1} \quad i = 1, 2,$$

$$a_{ij} = a_{ij}^{(0)} + a_{ij}^{(1)}\sqrt{-1} \quad i = 1, \cdots, \frac{r-1}{2}; j = 1, 2,$$

$$b_{ij} = b_{ij}^{(0)} + b_{ij}^{(1)}\sqrt{-1} \quad i = 1, \cdots, \frac{r-1}{2}; j = 1, 2,$$

where $a_i^{(j)}, a_{ij}^{(k)}$ and $b_{ij}^{(k)}$ are in $\mathbb{Z}_2$. Then the system of equations (24) ,(25) and (26) is equivalent to the following system of equations (27), (28) and (29):

(27) $\qquad (a_1^{(0)})^2 + (a_1^{(1)})^2 \equiv 0 \pmod{2\mathbb{Z}_2}$,

(28) $\qquad (a_2^{(0)})^2 + (a_2^{(1)})^2 \equiv 0 \pmod{2\mathbb{Z}_2}$,

(29) $\qquad \epsilon(a_1^{(0)}a_2^{(0)} + a_1^{(1)}a_2^{(1)}) + \epsilon(a_1^{(0)}a_2^{(1)} - a_1^{(1)}a_2^{(0)})\sqrt{-1}$

$$+ \sum_{i=1}^{\frac{r-1}{2}}(a_{i2}^{(0)}b_{i1}^{(0)} + a_{i1}^{(0)}b_{i2}^{(0)} + a_{i2}^{(1)}b_{i1}^{(1)} + a_{i1}^{(1)}b_{i2}^{(1)})$$

$$+ \sqrt{-1}\sum_{i=1}^{\frac{r-1}{2}}(a_{i2}^{(1)}b_{i1}^{(0)} - a_{i1}^{(1)}b_{i2}^{(0)} - a_{i2}^{(0)}b_{i1}^{(1)} + a_{i1}^{(0)}b_{i2}^{(1)}) \equiv 1 \pmod{2\mathbb{Z}_2}.$$

The equation (29) is equivalent to the following two equations (30) and (31):

(30) $\qquad \displaystyle\sum_{i=1}^{\frac{r-1}{2}}(a_{i2}^{(0)}b_{i1}^{(0)} + a_{i1}^{(0)}b_{i2}^{(0)} + a_{i2}^{(1)}b_{i1}^{(1)} + a_{i1}^{(1)}b_{i2}^{(1)})$

$$\equiv 1 + (a_1^{(0)}a_2^{(0)} + a_1^{(1)}a_2^{(1)}) \pmod{2\mathbb{Z}_2},$$

(31) $\qquad \displaystyle\sum_{i=2}^{\frac{r-1}{2}}(a_{i2}^{(1)}b_{i1}^{(0)} + a_{i1}^{(1)}b_{i2}^{(0)} + a_{i2}^{(0)}b_{i1}^{(1)} + a_{i1}^{(0)}b_{i2}^{(1)})$

$$\equiv a_1^{(0)}a_2^{(1)} + a_1^{(1)}a_2^{(0)} \pmod{2\mathbb{Z}_2}.$$

For any $a_i^{(j)}$ satisfying (27) and (28), we have $a_1^{(0)}a_2^{(0)} + a_1^{(1)}a_2^{(1)} \equiv a_1^{(0)}a_2^{(1)} + a_1^{(1)}a_2^{(0)}$. Let

$$A = \begin{pmatrix} \cdots & a_{i2}^{(0)}, a_{i1}^{(0)}, a_{i2}^{(1)}, a_{i1}^{(1)}, & \cdots \\ \cdots & a_{i2}^{(1)}, a_{i1}^{(1)}, a_{i2}^{(0)}, a_{i1}^{(0)}, & \cdots \end{pmatrix}$$

be a $(2, 2(r-1))$-matrix. Then (30) and (31) show that the rank of A is 2. Therefore for fixed $a_i^{(j)}, a_{ij}^{(k)}$ the number of the solutions of (30) and (31) is $2^{2(r-1)-2} = 2^{2r-4}$. On the other hand, the number of such matrices A of rank 2 is $2^{2(r-1)} - 2^{r-1}$, and the number of solutions satisfying (27) and (28) is 4.

Therefore we have $A_2(\mathcal{M}, H_2) = 2^{4(r-1)}(1 - 2^{-(r-1)})$. By Propositions 5.4 and 5.10

$$A_{2^3}(\mathcal{M}, H_2) = 2^{4(r-1)}A_{2^2}(\mathcal{M}, H_2) = 2^{8(r-1)}A_2(\mathcal{M}, H_2)$$

$$= 2^{12(r-1)}(1 - 2^{-(r-1)}).$$

Hence we have by Proposition 5.11

$$A_{2^3}(\mathcal{M}, \mathcal{M}) = A_{2^3}(\mathcal{M}, H_2) A_{2^3}(\mathcal{N}, \mathcal{N})$$
$$= 2^{12(r-1)}(1 - 2^{-(r-1)}) A_{2^3}(\mathcal{N}, \mathcal{N})$$

Therefore inductive computation gives

$$A_{2^3}(\mathcal{M}, \mathcal{M}) = 2^{\sum_{i=1}^{\frac{r-1}{2}} 24i} \prod_{i=1}^{\frac{r-1}{2}} (1 - 2^{-2i}) A_{2^3}(\langle \epsilon \rangle, \langle \epsilon \rangle)$$

It is easy to see $A_{2^3}(\langle \epsilon \rangle), \langle \epsilon \rangle)) = 2^4$. Thus we have $A_{2^3}(\mathcal{M}, \mathcal{M}) = 2^{3r^2+1} \prod_{i=1}^{\frac{r-1}{2}} (1 - 2^{-2i})$.

Since, by Proposition 5.1, $\beta_2(\mathcal{M}) = 2^{-3r^2} A_{2^3}(\mathcal{M}, \mathcal{M})$, we have

$$\beta_2(\mathcal{M}) = 2 \sum_{i=1}^{\frac{r-1}{2}} (1 - 2^{-2i}).$$

$\square$

Proposition 5.14. Let $\mathcal{M} = \langle \epsilon \rangle \perp \langle 1 \rangle \perp H_{r-2}$ with some even integer $r \geq 2$ and a unit ϵ in $\mathbf{Z}_2$. Then we have $\beta_2(\mathcal{M}) = 2 \prod_{i=1}^{\frac{r-2}{2}} (1 - 2^{-2i})$.

Proof. First we estimate $A_2(\mathcal{M}, H_2)$. Let

$$C = \begin{pmatrix} a_1 b_1 a_{11} b_{11} & \cdots & a_{\frac{r-2}{2},1} b_{\frac{r-2}{2},1} \\ a_2 b_2 a_{12} b_{12} & \cdots & a_{\frac{r-2}{2},2} b_{\frac{r-2}{2},2} \end{pmatrix}$$

be a $(r, 2)$-matrix over S_2 such that

$$(32) \qquad\qquad {}^t\overline{C} H_{\mathcal{M}} C \equiv H_2 \pmod{2S_2},$$

where $H_{\mathcal{M}} = \begin{pmatrix} \epsilon & & 0 \\ & 1 & \\ 0 & & H_{r-2} \end{pmatrix}$. Let $a_i = a_i^{(0)} + a_i^{(1)}\sqrt{-1}, b_i = b_i^{(0)} + b_i^{(1)}\sqrt{-1}$,
$a_{ij} = a_{ij}^{(0)} + a_{ij}^{(1)}\sqrt{-1}$ and $b_{ij} = b_{ij}^{(0)} + b_{ij}^{(1)}\sqrt{-1}$, where $a_i^{(j)}, a_{ij}^{(k)}, b_{ij}^{(k)}$ are in $\mathbf{Z}_2$. Then, equation (32) is equivalent to the following equations.

$$(33) \qquad\qquad (a_1^{(0)})^2 + (a_1^{(1)})^2 + (b_1^{(0)})^2 + (b_1^{(1)})^2 \equiv 0 \pmod{2\mathbf{Z}_2}$$
$$(34) \qquad\qquad (a_2^{(0)})^2 + (a_2^{(1)})^2 + (b_2^{(0)})^2 + (b_2^{(1)})^2 \equiv 0 \pmod{2\mathbf{Z}_2}$$

$$(35) \qquad \sum_{i=1}^{\frac{r-2}{2}} (a_{i2}^{(0)} b_{i1}^{(0)} + a_{i1}^{(0)} b_{i2}^{(0)} + a_{i2}^{(1)} b_{i1}^{(1)} + a_{i1}^{(1)} b_{i2}^{(1)})$$
$$\equiv 1 + a_1^{(0)} a_2^{(0)} + a_1^{(1)} a_2^{(1)} + b_1^{(0)} b_2^{(0)} + b_1^{(1)} b_2^{(1)} \pmod{2\mathbf{Z}_2}$$

$$(36) \qquad \sum_{i=1}^{\frac{r-2}{2}} (a_{i2}^{(1)} b_{i1}^{(0)} + a_{i1}^{(1)} b_{i2}^{(0)} + a_{i2}^{(0)} b_{i1}^{(1)} + a_{i1}^{(0)} b_{i2}^{(1)})$$
$$\equiv a_1^{(0)} a_2^{(1)} + a_1^{(1)} a_2^{(0)} + b_1^{(0)} b_2^{(1)} + b_1^{(1)} b_2^{(0)} \pmod{2\mathbf{Z}_2}.$$

For any $a_i^{(j)}$ and $b_i^{(j)}$, $i = 1, 2; j = 0, 1$, satisfying (33) and (34) we have $a_1^{(0)}a_2^{(1)} + a_1^{(1)}a_2^{(0)} + b_1^{(0)}b_2^{(1)} + b_1^{(1)}b_2^{(0)} \equiv a_1^{(0)}a_2^{(0)} + a_1^{(1)}a_2^{(1)} + b_1^{(0)}b_2^{(0)} + b_1^{(1)}b_2^{(1)}$ $(mod\ 2\mathbf{Z}_2)$.

Therefore, the $(2, 2(r-2))$-matrix $A = \begin{pmatrix} \cdots & a_{i2}^{(0)}a_{i1}^{(0)}a_{i2}^{(1)}a_{i1}^{(1)} & \cdots \\ \cdots & a_{i2}^{(1)}a_{i1}^{(1)}a_{i2}^{(0)}a_{i1}^{(0)} & \cdots \end{pmatrix}$ is of rank 2.

For a fixed A, and $a_i^{(j)}, b_i^{(j)}$ satisfying (33) and (34), the number of solutions satisfying (35) and (36) is 2^{2r-6}. The number of such matrices A of rank 2 is $2^{2(r-2)} - 2^{r-2}$ and the number of the solutions satisfying (33) and (34) is 2^6. Hence we have $A_2(M, H_2) = 2^{4r-4}(1 - 2^{-(r-2)})$ from which $A_{2^3}(M, H_2) = 2^{12(r-1)}(1 - 2^{-(r-2)})$ follows by Propositions 5.4 and 5.10.

Let $\mathcal{N} = \langle \epsilon \rangle \perp \langle 1 \rangle \perp H_{r-4}$. Then Proposition 5.11 gives

$$A_{2^3}(M, M) = A_{2^3}(M, H_2)A_{2^3}(\mathcal{N}, \mathcal{N})$$
$$= 2^{12(r-1)}(1 - 2^{-(r-2)})A_{2^3}(\mathcal{N}, \mathcal{N}).$$

By inductive computation we have

$$A_{2^3}(M, M) = 2^{3r^2-12} \sum_{i=1}^{\frac{r-2}{2}} (1 - 2^{-2i})A_{2^3}\left(\begin{pmatrix} \epsilon & 0 \\ 0 & 1 \end{pmatrix}, \begin{pmatrix} \epsilon & 0 \\ 0 & 1 \end{pmatrix}\right).$$

Let $C = \begin{pmatrix} a \\ b \end{pmatrix}$, $a, b \in S_2$, such that

(37) $${}^t\overline{C}\begin{pmatrix} \epsilon & 0 \\ 0 & 1 \end{pmatrix} C \equiv \epsilon \ (mod\ 2^3 S_2).$$

An argument similar to that on page 28 shows that there exists a $(2, 2)$-matrix $A = \begin{pmatrix} a & * \\ b & * \end{pmatrix}$ satisfying ${}^t\overline{A}\begin{pmatrix} \epsilon & 0 \\ 0 & 1 \end{pmatrix} A \equiv \begin{pmatrix} \epsilon & 0 \\ 0 & 1 \end{pmatrix} (mod\ 2^3 S_2)$. Let $X = (x_{ij})$ be a $(2, 2)$-matrix such that $\begin{pmatrix} x_{11} \\ x_{21} \end{pmatrix} \equiv \begin{pmatrix} a \\ b \end{pmatrix} (mod\ 2^3 S_2)$ and ${}^t\overline{X}\begin{pmatrix} \epsilon & 0 \\ 0 & 1 \end{pmatrix} X \equiv \begin{pmatrix} \epsilon & 0 \\ 0 & 1 \end{pmatrix}$ $(mod\ 2^3 S_2)$. Then we have $X \equiv A \begin{pmatrix} 1 & c \\ 0 & x \end{pmatrix}$ with some $c, x \in S_2$. Since ${}^t\overline{X}\begin{pmatrix} \epsilon & 0 \\ 0 & 1 \end{pmatrix}$ $X \equiv \begin{pmatrix} 1 & 0 \\ 0 & \overline{x} \end{pmatrix}\begin{pmatrix} \epsilon & 0 \\ 0 & 1 \end{pmatrix}\begin{pmatrix} 1 & c \\ 0 & x \end{pmatrix} = \begin{pmatrix} 1 & c \\ \epsilon\overline{c} & \overline{x}x \end{pmatrix}$ and ${}^t\overline{X}\begin{pmatrix} \epsilon & 0 \\ 0 & 1 \end{pmatrix} X \equiv \begin{pmatrix} \epsilon & 0 \\ 0 & 1 \end{pmatrix}$, we have $c \equiv 0 \ (mod\ 2^3 S_2)$.

Therefore

$$A_{2^3}\left(\begin{pmatrix} \epsilon & 0 \\ 0 & 1 \end{pmatrix}, \begin{pmatrix} \epsilon & 0 \\ 0 & 1 \end{pmatrix}\right) = A_{2^3}\left(\begin{pmatrix} \epsilon & 0 \\ 0 & 1 \end{pmatrix}, \langle \epsilon \rangle\right)A_{2^3}(\langle 1 \rangle, \langle 1 \rangle) = 2^4 A_{2^3}\left(\begin{pmatrix} \epsilon & 0 \\ 0 & 1 \end{pmatrix}, \langle \epsilon \rangle\right).$$

Now we estimate $A_{2^3}\left(\begin{pmatrix} \epsilon & 0 \\ 0 & 1 \end{pmatrix}, \langle \epsilon \rangle\right)$. Let $a = a_0 + a_1\sqrt{-1}$ and $b = b_0 + b_1\sqrt{-1}$ with $a_i, b_i \in \mathbf{Z}_2$. Equation (37) is equivalent to

(38) $$\epsilon a_0^2 + \epsilon a_1^2 + b_0^2 + b_1^2 \equiv \epsilon \ (mod\ 2^3\mathbf{Z}_2)$$

Let $T(a)$ be the number of the solutions of $x^2 + y^2 \equiv a \ (mod\ 2^3\mathbf{Z}_2)$. We have $T(0) = T(4) = 2^3$, $T(1) = T(2) = T(5) = 2^4$ and $T(3) = T(7) = T(6) = 0$.

Then the number of the solutions of (38) is $2(T(0)T(1) + T(2)T(7) + T(3)T(6) + T(4)T(5)) = 2^9$ if $\epsilon = 1$, and $T(0)T(1) + T(1)T(6) + T(2)T(3) + T(3)T(0) + T(4)T(5) + T(5)T(2) + T(6)T(7) + T(7)T(4) = 2^9$ if $\epsilon = 3$. Hence $A_{2^3}(\begin{pmatrix} \epsilon & 0 \\ 0 & 1 \end{pmatrix}, \langle \epsilon \rangle) = 2^9$

and thus, $A_{2^3}(\begin{pmatrix} \epsilon & 0 \\ 0 & 1 \end{pmatrix}, \begin{pmatrix} \epsilon & 0 \\ 0 & 1 \end{pmatrix}) = 2^{13}$ and $A_{2^3}(M, M) = 2^{3r^2+1} \prod_{i=1}^{\frac{r-2}{2}} (1 - 2^{-2i})$. Since $\beta_2(M) = 2^{-3r^2} A_{2^3}(M, M)$ by Proposition 5.1, we have the proposition. $\square$

§6. List of quadratic lattices over 2-adic integers and their local densities.

The classifiction of quadratic lattices over 2-adic integers has been completed by O'Meara (see [8]). Local densities of quadratic lattices at the prime 2 have been obtained by Watson (see [16]).

In this section we give the local densities of quadratic $\mathbf{Z}_2$-lattices N satisfying the following conditions.

(i) rank$_{\mathbf{Z}_2} N = n$ and $dN = 2^{n-\rho}$ with some integer ρ such that $0 \le \rho \le n$

(ii) $N = N_u \perp N_t$ with some sublattices N_u and N_t of N such that N_u is unimodular of rank ρ and N_t is $2\mathbf{Z}_2$-modular of rank $n - \rho$.

Notation:

$$H_i = \begin{pmatrix} 0 & 1 \\ 1 & 0 \end{pmatrix} \perp \cdots \perp \begin{pmatrix} 0 & 1 \\ 1 & 0 \end{pmatrix} \text{ of rank } i \ge 0,$$

$$A_i = \begin{pmatrix} 2 & 1 \\ 1 & 2 \end{pmatrix} \perp \begin{pmatrix} 0 & 1 \\ 1 & 0 \end{pmatrix} \perp \cdots \perp \begin{pmatrix} 0 & 1 \\ 1 & 0 \end{pmatrix} \text{ of rank } i \ge 2,$$

$$P_2(k) = \prod_{i=1}^{k} (1 - 2^{-2i}), P_2(0) = 1,$$

the small letters a, b, c and d denote units in $\mathbf{Z}_2$, and $2H_i, 2\langle a \rangle$, etc., denote the lattices $H_i, \langle a \rangle$, etc., (respectively), scaled by 2.

Let N be a lattice over $\mathbf{Z}_2$ satisfying the conditons (i) and (ii) given above, then N is one of the following 14 types of lattices

1) $n =$ even, $\rho =$ odd, $\langle a \rangle \perp H_{\rho-1} \perp 2\langle b \rangle \perp 2H_{n-\rho-1}$, with $ab \equiv (-1)^{\frac{n-2}{2}}$ modulo square. The number of classes of this type is 4.

2) $n \equiv 0 \pmod 4$, $\rho =$ even ≥ 0, $H_\rho \perp 2H_{n-\rho}$. There is exactly one class of this type.

3) $n \equiv 0 \pmod 4$, $\rho =$ even ≥ 2, $n - \rho \ge 2$ $A_\rho \perp 2 A_{n-\rho}$. There is exactly one class of this type.

4) $n =$ even, $\rho =$ even, $n - \rho \ge 2, H_\rho \perp 2\{\langle a \rangle \perp \langle b \rangle \perp H_{n-\rho-2}\}$, with $a \equiv b \pmod 4$ for $n \equiv 2 \pmod 4$, and $a \not\equiv b \pmod 4$ for $n \equiv 0 \pmod 4$. The number of classes of this type is 2.

5) $n \equiv 0 \pmod 4, \rho =$ even, $n - \rho \ge 4$, $H_\rho \perp 2\{\langle a \rangle \perp \langle b \rangle \perp A_{n-\rho-2}\}$ with $a \not\equiv b \pmod 4$. The number of classes of this type is 1.

6) n = even, ρ = even ≥ 2, $\langle a \rangle \perp \langle b \rangle \perp H_{\rho-2} \perp 2H_{n-\rho}$, with $ab \equiv -1$ modulo square if $n \equiv 0 \ (mod\ 4)$ and $ab \equiv 1$ modulo square if $n \equiv 2 \ (mod\ 4)$. The number of classes of this type is 2.

7) $n \equiv 0 \ (mod\ 4)$, ρ = even ≥ 4, $\langle a \rangle \perp \langle b \rangle \perp A_{\rho-2} \perp 2H_{n-\rho}$, with $a \not\equiv b \ (mod\ 4)$. The number of classes of this type is 1.

8) n = even, ρ = even ≥ 2, $n - \rho \geq 2$, $\langle a \rangle \perp \langle b \rangle \perp H_{\rho-2} \perp 2\{\langle c \rangle \perp \langle d \rangle \perp H_{n-\rho-2}\}$, with $abcd \equiv (-1)^{\frac{n}{2}}$ modulo square. The number of classes of this type is 16.

9) n = odd, ρ = odd ≥ 1, $\langle a \rangle \perp H_{\rho-1} \perp 2H_{n-\rho}$, with $a \equiv (-1)^{\frac{n-1}{2}}$ modulo square. There is exactly one class of this type.

10) n = odd, ρ = odd ≥ 3, $\langle a \rangle \perp A_{\rho-1} \perp 2H_{n-\rho}$, with $3a \equiv (-1)^{\frac{n-3}{2}}$ modulo square. There is exactly one class of this type.

11) n = odd, ρ = odd ≥ 1, $\langle a \rangle \perp H_{\rho-1} \perp 2\{\langle b \rangle \perp \langle c \rangle \perp H_{n-\rho-2}\}$, with $abc \equiv (-1)^{\frac{n-3}{2}}$ modulo square. The number of classes of this type is 8.

12) n = odd, ρ = even $\leq n - 1$, $H_\rho \perp 2\{\langle a \rangle \perp H_{n-\rho-1}\}$, with $a \equiv (-1)^{\frac{n-1}{2}}$ modulo square. There is exactly one class of this type.

13) n = odd, ρ = even $\leq n - 3$, $H_\rho \perp 2\{\langle a \rangle \perp A_{n-\rho-1}\}$, with $a \equiv 3(-1)^{\frac{n-3}{2}}$ modulo square. There is exactly one class of this type.

14) n = odd, ρ = even ≥ 2, $\rho \leq n - 1$, $\langle a \rangle \perp \langle b \rangle \perp H_{\rho-2} \perp 2\{\langle c \rangle \perp H_{n-\rho-1}\}$ with $abc \equiv (-1)^{\frac{n-3}{2}}$ modulo square. The number of classes of this type is 8.

Local densities of the lattices given above are as follows:

N = **type 1)**, then

$$\alpha_2(N) = 2^{\frac{1}{2}n^2 + \frac{1}{2}n - n\rho + \frac{1}{2}(\rho^2 - \rho) + 2} P_2(\frac{\rho - 1}{2}) P_2(\frac{n - \rho - 1}{2}).$$

N = **type 2) or 3)**, then

$$\alpha_2(N) = \frac{2^{\frac{1}{2}n^2 + \frac{3}{2}n - n\rho + \frac{1}{2}(\rho^2 - \rho) + 1} P_2(\frac{\rho}{2}) P_2(\frac{n-\rho}{2})}{(1 + \delta 2^{-\frac{\rho}{2}})(1 + \delta 2^{-\frac{1}{2}(n-\rho)})},$$

where $\delta = 1$ for type 2) and $\delta = -1$ for type 3).

N = **type 4) or 5)**, then

$$\alpha_2(N) = \frac{2^{\frac{1}{2}n^2 + \frac{1}{2}n - n\rho + \frac{1}{2}(\rho^2 + \rho) - 2} P_2(\frac{\rho}{2}) P_2(\frac{n-\rho-2}{2})}{1 + \delta 2^{-\frac{1}{2}(n-\rho-2)}},$$

where $\delta = 0$ or 1 for type 4) and $\delta = -1$ for type 5).

N = **type 6) or 7)**, then

$$\alpha_2(N) = \frac{2^{\frac{1}{2}n^2 + \frac{3}{2}n - n\rho + \frac{1}{2}(\rho^2 - 3\rho) + 2} P_2(\frac{\rho-2}{2}) P_2(\frac{n-\rho}{2})}{1 + \delta 2^{-\frac{1}{2}(\rho-2)}},$$

where $\delta = 0$ or 1 for type 6) and $\delta = -1$ for type 7).

N = **type 8)**, then

$$\alpha_2(N) = 2^{\frac{1}{2}n^2+\frac{1}{2}n-n\rho+\frac{1}{2}(\rho^2+\rho)}P_2(\frac{\rho-2}{2})P_2(\frac{n-\rho-2}{2}).$$

N = **type 9) or 10)**, then

$$\alpha_2(N) = \frac{2^{\frac{1}{2}n^2+\frac{3}{2}n-n\rho+\frac{1}{2}(\rho^2-3\rho)+2}P_2(\frac{\rho-1}{2})P_2(\frac{n-\rho}{2})}{1+\delta 2^{-\frac{1}{2}(\rho-1)}},$$

where $\delta = 1$ for type 9) and $\delta = -1$ for type 10).

N = **type 11)**, then

$$\alpha_2(N) = 2^{\frac{1}{2}n^2+\frac{1}{2}n-n\rho+\frac{1}{2}(\rho^2-\rho)+2}P_2(\frac{\rho-1}{2})P_2(\frac{n-\rho-2}{2}).$$

N = **type 12) or 13)**, then

$$\alpha_2(N) = \frac{2^{\frac{1}{2}n^2+\frac{1}{2}n-n\rho+\frac{1}{2}(\rho^2+\rho)+2}P_2(\frac{\rho}{2})P_2(\frac{n-\rho-1}{2})}{1+\delta 2^{-\frac{1}{2}(n-\rho-1)}},$$

whre $\delta = 1$ for type 12) and $\delta = -1$ for type 13).

N = **type 14)**, then

$$\alpha_2(N) = 2^{\frac{1}{2}n^2+\frac{1}{2}n-n\rho+\frac{1}{2}(\rho^2-\rho)+2}P_2(\frac{\rho-2}{2})P_2(\frac{n-\rho-1}{2}).$$

For fixed n and ρ, the following table gives the number of the classes of lattices satisfying conditions (i) and (ii) given at the beginning of this section.

n	ρ	number of classes
$n = $ even	$\rho = $ odd	4
$n = 4$	$\rho = 0$	4
	$\rho = 2$	22
	$\rho = 4$	4
$n \equiv 0 \, (mod \, 4), \ n \geq 8$	$\rho = 0$	4
	$\rho = 2$	23
	$4 \leq \rho \leq n - 4, \rho = $ even	24
	$\rho = n - 2$	23
	$\rho = n$	4
$n \equiv 2 \, (mod \, 4)$	$\rho = 0$	2
	$2 \leq \rho \leq n - 2, \rho = $ even	20
	$\rho = n$	2
$n = 1$	$\rho = 0$	1
	$\rho = 1$	1
$n = $ odd, $n \geq 3$	$\rho = 0$	2
	$\rho = 1$	9
	$2 \leq \rho \leq n - 2$	10
	$\rho = n - 1$	9
	$\rho = n$	2

CHAPTER IV
ESTIMATIONS

In this chapter we estimate $\omega_{R(2)}/\omega(L)$ (in §7), $\omega_{R(q)}/\omega(L)$ (for $q \neq 2$) (in §8) and $\omega_{IR(q)}/\omega(L)$ (in §9). First we give some notation and some estimations of functions we need in this chapter.

Notation:

$$X(N_0, \mathcal{N}_1) = |I(V_0'(N_0), V_1'(\mathcal{N}_1))|\omega(N_0)\omega(\mathcal{N}_1)/\omega(L) \text{ (if } q = 2 \text{ then } \mathcal{N}_1 = N_1).$$

$$P_p(k) = \prod_{i=1}^{k}(1 - p^{-2k}) \text{ for any prime p.}$$

$$\varsigma(k) = \sum_{n=1}^{\infty}\frac{1}{n^k} = \prod_{p=prime}(1 - p^{-k})^{-1}.$$

(39)
$$Y(m, r, 2) = \frac{2^{\frac{1}{2}r(m-r-1)}\pi^{\frac{1}{2}r(m-r)}\prod_{i=1}^{m_0}\Gamma(\frac{i}{2})\prod_{i=1}^{r}\Gamma(\frac{i}{2})}{\prod_{i=1}^{m}\Gamma(\frac{i}{2})}.$$

(40)
$$Y(m, r, q) = 2^{\frac{1}{4}r(q-1)(2m-rq-3)}\pi^{\frac{1}{4}r(q-1)(2m-rq-1)}$$
$$\cdot q^{\frac{1}{4}r(2m-rq-1)}(\prod_{j=1}^{r}(j-1)!)^{\frac{q-1}{2}}\{m_0!(m_0+2)!\cdots(m-2)!\}^{-1},$$

where $m_0 = m - r(q - 1)$.

Local densities at nondyadic primes.

Let N be an integral $\mathbf{Z}$-lattice of rank n. Then

(41)
$$\alpha_p(N) = P_p(\frac{n-1}{2}) \text{ for } n = odd \text{ and } p \nmid 2dN,$$

(42)
$$\alpha_p(N) = (1 - (\frac{(-1)^{\frac{n}{2}}dN}{p})p^{-\frac{n}{2}})P_p(\frac{n-2}{2}) \text{ for } n = even$$

and $p \nmid 2dN$ (see Hilfssatz 12 in [14]).

If q is an odd prime and

$$N_q \cong \underbrace{\langle 1 \rangle \perp \cdots \perp \langle 1 \rangle \perp \langle \epsilon \rangle}_{\rho \; copies} \perp \langle q \rangle \perp \cdots \perp \langle q\epsilon \rangle$$

then we have the following (43) - (46) (see 6.4 in [16]).

$$(43) \qquad \alpha_q(N) = \frac{2q^{\frac{1}{2}(n-\rho)(n-\rho+1)}P_q(\frac{\rho-1}{2})P_q(\frac{n-\rho}{2})}{1 + \delta q^{-\frac{1}{2}(n-\rho)}}$$

for n = odd and ρ = odd, where $\delta = (\frac{(-1)^{\frac{n-\rho}{2}}\epsilon}{q})$.

$$(44) \qquad \alpha_q(N) = \frac{2q^{\frac{1}{2}(n-\rho)(n-\rho+1)}P_q(\frac{\rho}{2})P_q(\frac{n-\rho-1}{2})}{1 + \delta q^{-\frac{1}{2}(n-\rho-1)}}$$

for n = odd and ρ = even, where

$$\delta = (\frac{(-1)^{\frac{n-\rho-1}{2}}\epsilon}{q}).$$

$$(45) \qquad \alpha_q(N) = 2q^{\frac{1}{2}(n-\rho)(n-\rho+1)}P_q(\frac{\rho-1}{2})P_q(\frac{n-\rho-1}{2})$$

for n = even and ρ = odd.

$$(46) \qquad \alpha_q(N) = \frac{2q^{\frac{1}{2}(n-\rho)(n-\rho+1)}P_q(\frac{\rho}{2})P_q(\frac{n-\rho}{2})}{(1 + \delta_1 q^{-\frac{\rho}{2}})(1 + \delta_2 q^{-\frac{n-\rho}{2}})}$$

for n = even and ρ = even, where

$$\delta_1 = (\frac{(-1)^{\frac{\rho}{2}}\epsilon}{q}) \text{ and } \delta_2 = (\frac{(-1)^{\frac{n-\rho}{2}}\epsilon}{q}).$$

Gamma function

$$(47) \qquad \Gamma(x)\Gamma(x + \frac{1}{2}) = 2^{1-2x}\pi^{\frac{1}{2}}\Gamma(2x),$$

$$(48) \qquad \Gamma(x) = \sqrt{2\pi}\, x^{x-\frac{1}{2}}e^{-x}e^{\eta(x)}, \text{ where } 0 < \eta(x) < \frac{1}{8x},$$

$$(49) \qquad \Gamma(n) = (n - 1)! \text{ for integer } n \geq 1,$$

$$(50) \qquad \varsigma(2) = \frac{\pi^2}{6}, \; \varsigma(k) < 1.01 \text{ if } k \geq 7,$$

(51)
$$\prod_p P_p(k)^{-1} < \varsigma(2)e^{\frac{1}{6}} \text{ for any integer } k \geq 2$$

(52)
$$\prod_{p \neq 2}(1 - (\frac{-1}{p})p^{-1})^{-1} = \frac{\pi}{4}, \prod_{p \neq 2}(1 - (\frac{-2}{p})p^{-1}) = \frac{\pi}{2\sqrt{2}},$$

(53)
$$\prod_p \prod_{i=3}^r (1 - p^{-i})^{-1} < e^{\frac{1}{2}} \text{ for any integer } r \geq 3,$$

(54)
$$\prod_{p \neq 2}(1 - (\frac{-1}{p})p^{-i}) < (1 - 3^{-i})^{-1},$$

(55)
$$\prod_{\substack{p \neq 2 \\ p \neq q}}(1 - (\frac{-q}{p})p^{-1})^{-1} < 0.36q \text{ for } q \geq 3$$

(see 5.7 in [16]).

The order of $I(V_0'(N_0), V_1'(N_1))$. (See [4], pages xi and xii)

If $q = 2$ and $V_0'(N_0) \cong V_1'(N_1) \cong \begin{pmatrix} 0 & 1 \\ 1 & 0 \end{pmatrix} \perp \cdots \perp \begin{pmatrix} 0 & 1 \\ 1 & 0 \end{pmatrix}$ then

(56)
$$|I(V_0'(N_0), V_1'(N_1))| = 2^{\frac{1}{2}\rho(\rho+1)} \prod_{i=1}^{\frac{\rho}{2}}(1 - 2^{-2i}).$$

If $q = 2$ and $V_0'(N_0) \cong V_1'(N_1) \cong \begin{pmatrix} 0 & 1 \\ 1 & 0 \end{pmatrix} \perp \cdots \perp \begin{pmatrix} 0 & 1 \\ 1 & 0 \end{pmatrix} \perp \langle 1 \rangle$ then

(57)
$$|I(V_0'(N_0), V_1'(N_1))| = 2^{\frac{1}{2}\rho(\rho-1)} \prod_{i=1}^{\frac{\rho-1}{2}}(1 - 2^{-2i}).$$

If $q = 2$ and $V_0'(N_0) \cong V_1'(N_1) \cong \begin{pmatrix} 0 & 1 \\ 1 & 0 \end{pmatrix} \perp \cdots \perp \begin{pmatrix} 0 & 1 \\ 1 & 0 \end{pmatrix} \perp \langle 1 \rangle \perp \langle 1 \rangle$ then

(58)
$$|I(V_0'(N_0), V_1'(N_1))| = 2^{\frac{1}{2}\rho(\rho-1)} \prod_{i=1}^{\frac{\rho-2}{2}}(1 - 2^{-2i}).$$

If $q = $ an odd prime then

(59)
$$|I(V_0'(N_0), V_1'(N_1))| = 2q^{\frac{1}{2}\rho(\rho-1)} P_q(\frac{\rho-1}{2})$$

for $\rho = $ odd and

(60)
$$|I(V_0'(N_0), V_1'(N_1))| = 2q^{\frac{1}{2}\rho(\rho-1)} P_q(\frac{\rho}{2})/(1 + \delta q^{-\frac{\rho}{2}})$$

for $\rho = $ even, where $\delta = (\frac{(-1)^{\frac{\rho}{2}}dV_0'(N_0)}{q})$.

§7. **Estimation of** $\omega_{R(2)}/\omega(L)$.

In this section we give an upper bound for $\omega_{R(2)}/\omega(L)$. Let N_0 and N_1 be the lattices such that the pair of genera (G_{N_0}, G_{N_1}) is in $G(2, r, \rho)$, where $1 \leq r \leq [\frac{m}{2}]$ and $0 \leq \rho \leq r$. First we evaluate

$$|I(V_0'(N_0), V_1'(N_1))|\omega(N_0)\omega(N_1)/\omega(L) = X(N_0, N_1).$$

Proposition 7.1. Let $r = 1$ and $m \geq 30$ then we have the followings.

  i) If L is odd unimodular then

$$X(N_0, N_1) < 1.01 \cdot 2^{\frac{m}{2}-2} \pi^{\frac{m}{2}} \Gamma(\frac{m}{2})^{-1}.$$

  ii) If L is even unimodular then

$$X(N_0, N_1) < 1.01 \cdot 2^{\frac{3}{2}m-4} \pi^{\frac{m}{2}} \Gamma(\frac{m}{2})^{-1}.$$

Proof. Since $r = 1$ we have $\rho = 0$ or 1, $\omega(N_1) = \frac{1}{2}$, and $|I(V_0'(N_0), V_1'(N_1))| = 1$. Therefore by the mass formula (1) we have

$$X(N_0, N_1) = \frac{1}{2} \cdot \frac{dN_0^{\frac{m}{2}}}{\alpha_2(N_0)} \cdot \alpha_2(L) \cdot \prod_{p \neq 2} \frac{\alpha_p(L)}{\alpha_p(N_0)} \cdot \pi^{\frac{m}{2}} \Gamma(\frac{m}{2})^{-1}.$$

If $m = odd$ then by (41) and (42)

$$\prod_{p \neq 2} \frac{\alpha_p(L)}{\alpha_p(N_0)} = \prod_{p \neq 2} \frac{P_p(\frac{m-1}{2})}{(1 - \delta p^{-\frac{m-1}{2}})P_p(\frac{m-3}{2})}$$
$$= \prod_{p \neq 2} (1 - \delta p^{-\frac{m-1}{2}})^{-1}(1 - p^{-(m-1)}),$$

where $\delta = (\frac{(-1)^{\frac{m-1}{2}}dN_0}{p})$.

Therefore we have $\prod_{p \neq 2} \frac{\alpha_p(L)}{\alpha_p(N_0)} < \prod_{p \neq 2}((1 - p^{-\frac{m-1}{2}})^{-1} = (1 - 2^{-\frac{m-1}{2}})\varsigma(\frac{m-1}{2})$. We also have $\alpha_2(L) = 2^2 P_2(\frac{m-1}{2})/(1 + \delta 2^{-\frac{m-1}{2}})$, where $\delta = \pm 1$ (see §6 type 9) or 10) with $\rho = n = m$).

  a) If $\rho = 0$, then $(N_0)_2$ is of type 2), 4) or 5) in §6, and $dN_0 = 2^{m-1}$. Therefore if N_0 is of type 2) then

$$\frac{dN_0^{\frac{m}{2}}}{\alpha_2(N_0)} \cdot \alpha_2(L) = 2^{-m+3}(1 + 2^{-\frac{m-1}{2}})/(1 + \delta 2^{-\frac{m-1}{2}}),$$

if N_0 is of type 4) or 5) then

$$\frac{dN_0^{\frac{m}{2}}}{\alpha_2(N_0)} \cdot \alpha_2(L) = 2^4(1 + \delta_0 2^{-\frac{m-3}{2}})(1 - 2^{-(m-1)})/(1 + \delta 2^{-\frac{m-1}{2}}),$$

where $\delta_0 = 0$ or ± 1 and $\delta = \pm 1$. Hence, if N_0 is of type 2), then

$$X(N_0, N_1) < 1.01 \cdot 2^{-m+2} \pi^{\frac{m}{2}} \Gamma(\frac{m}{2})^{-1},$$

if N_0 is of type 4) or 5), then

$$X(N_0, N_1) < 1.01 \cdot 2^3 \pi^{\frac{m}{2}} \Gamma(\frac{m}{2})^{-1}$$

b) If $\rho = 1$ then $(N_0)_2$ is of type 1) and

$$\frac{dN_0^{\frac{m}{2}}}{\alpha_2(N_0)} \cdot \alpha_2(L) = 2^{\frac{m}{2}-1}(1 - 2^{-(m-1)})/(1 + \delta 2^{-\frac{m-1}{2}}).$$

Hence we have

$$X(N_0, N_1) < 1.01 \cdot 2^{\frac{m}{2}-2} \pi^{\frac{m}{2}} \Gamma(\frac{m}{2})^{-1}$$

If m = even then by (41) and (42)

$$\prod_{p \neq 2} \frac{\alpha_p(L)}{\alpha_p(N_0)} = \prod_{p \neq 2} \frac{(1 - \delta p^{-\frac{m}{2}}) P_p(\frac{m-2}{2})}{P_p(\frac{m-2}{2})} = \prod_{p \neq 2}(1 - \delta p^{-\frac{m}{2}}),$$

where $\delta = (\frac{(-1)^{\frac{m}{2}}}{p})$. Hence by (54) $\prod_{p \neq 2} \frac{\alpha_p(L)}{\alpha_p(N_0)} < (1 - 3^{-\frac{m}{2}})^{-1}$.

Also we have the followings:

If L is odd then

$$\alpha_2(L) = 2^2 P_2(\frac{m-2}{2})/(1 + \delta 2^{-\frac{m-2}{2}}),$$

where $\delta = 0, \pm 1$ (see §6 type 6) or 7) with $\rho = n = m$). If L is even then

$$\alpha_2(L) = 2^m P_2(\frac{m}{2})/(1 + 2^{-\frac{m}{2}}),$$

(see §6 type 2) with $\rho = n = m$).

a) If $\rho = 0$ then $(N_0)_2$ is of type 12) or 13) in §6 and $dN_0 = 2^{m-1}$. Hence we have if L is odd then

$$\frac{dN_0^{\frac{m}{2}}}{\alpha_2(N_0)} \cdot \alpha_2(L) = \frac{1 + \delta_0 1^{-\frac{m-2}{2}}}{1 + \delta 2^{-\frac{m-2}{2}}},$$

where $\delta_0 = \pm 1$ and $\delta = 0, \pm 1$, and if L is even then

$$\frac{dN_0^{\frac{m}{2}}}{\alpha_2(N_0)} \cdot \alpha_2(L) = \frac{2^{m-2}(1 + \delta_0 2^{-\frac{m-2}{2}})(1 - 2^{-m})}{1 + 2^{-\frac{m}{2}}},$$

where $\delta_0 = \pm 1$.

Hence we have

$$X(N_0, N_1) < \frac{1}{2} \cdot 1.01 \cdot \pi^{\frac{m}{2}} \Gamma(\frac{m}{2})^{-1},$$

for L odd and

$$X(N_0, N_1) < 1.01 \cdot 2^{m-3} \pi^{\frac{m}{2}} \Gamma(\frac{m}{2})^{-1}$$

for L even.

b) If $\rho = 1$ then $(N_0)_2$ is of type 9) or 11) and $dN_0 = 2^{m-2}$. Hence, if L is odd and $(N_0)_2$ is of type 9) then

$$\frac{dN_0^{\frac{m}{2}}}{\alpha_2(N_0)} \cdot \alpha_2(L) = 2^{-\frac{m}{2}+2}/(1 + \delta 2^{-\frac{m-2}{2}}),$$

and if L is odd and $(N_0)_2$ is of type 11) then

$$\frac{dN_0^{\frac{m}{2}}}{\alpha_2(N_0)} \cdot \alpha_2(L) = 2^{\frac{m}{2}-1}(1 - 2^{-(m-2)})(1 + \delta 2^{-\frac{m-2}{2}}),$$

if L is even and $(N_0)_2$ is of type 9) then

$$\frac{dN_0^{\frac{m}{2}}}{\alpha_2(N_0)} \cdot \alpha_2(L) = 2^{\frac{m}{2}}(1 - 2^{-m})/(1 + 2^{-\frac{m}{2}}),$$

and if L is even and $(N_0)_2$ is of type 11) then

$$\frac{dN_0^{\frac{m}{2}}}{\alpha_2(N_0)} \cdot \alpha_2(L) = 2^{\frac{3}{2}m-3}(1 - 2^{-m})(1 - 2^{-(m-2)})/(1 + 2^{-\frac{m}{2}}),$$

Hence if L is odd and $(N_0)_2$ is of type 9) then

$$X(N_0, N_1) < 1.01 \cdot 2^{-\frac{m}{2}+1} \pi^{\frac{m}{2}} \Gamma(\frac{m}{2})^{-1},$$

if L is odd and $(N_0)_2$ is of type 11) then

$$X(N_0, N_1) < 1.01 \cdot 2^{\frac{m}{2}-2} \pi^{\frac{m}{2}} \Gamma(\frac{m}{2})^{-1},$$

if L is even and $(N_0)_2$ is of type 9) then

$$X(N_0, N_1) < 1.01 \cdot 2^{\frac{m}{2}-1} \pi^{\frac{m}{2}} \Gamma(\frac{m}{2})^{-1},$$

if L is even and $(N_0)_2$ is of type 11) then

$$X(N_0, N_1) < 1.01 \cdot 2^{\frac{3}{2}m-4} \pi^{\frac{m}{2}} \Gamma(\frac{m}{2})^{-1}. \qquad \square$$

Now we assume $r \geq 2$. Then we have

(61) $$X(N_0, N_1) = \frac{2|I(V_0'(N_0), V_1'(N_1)|dN_0^{\frac{m_0+1}{2}} dN_1^{\frac{r+1}{2}}}{\alpha_2(N_0)\alpha_2(N_1)} \cdot \alpha_2(L)$$

$$\cdot \prod_{p \neq 2} \frac{\alpha_p(L)}{\alpha_p(N_0)\alpha_p(N_1)} \cdot \pi^{\frac{1}{2}r(m-r)} \cdot \frac{\prod_{i=1}^{m_0} \Gamma(\frac{i}{2}) \prod_{i=1}^{r} \Gamma(\frac{i}{2})}{\prod_{i=1}^{m} \Gamma(\frac{i}{2})}.$$

Define $F(\rho) = \dfrac{2|I(V_0'(N_0),V_1'(N_1)|dN_0^{\frac{m_0+1}{2}}dN_1^{\frac{r+1}{2}}}{\alpha_2(N_0)\alpha_2(N_1)}$. Then other factors in (61) do not depend on ρ.

The following is the list of possible pairs of N_0 and N_1 such that (G_{N_0}, G_{N_1}) $\in G(2,r,\rho)$ with $r \geq 2$, along with the corresponding value $F(\rho)$. In this table $\delta_i \in \{0, 1, -1\}$ depends on the type of $N_i, i = 0, 1$ (see §6). Number of the pairs for fixed m, r and ρ is also given.

$\underline{m = odd,\ r = odd\ (\geq 3)\ and\ \rho = odd}$

i) N_0 is type 1) and N_1 is type 9) or 10),

$$F(\rho) = \frac{2^{-r-4+\frac{1}{2}\rho(m-\rho+1)}(1 + \delta_1 2^{-\frac{\rho-1}{2}})}{P_2(\frac{\rho-1}{2})P_2(\frac{m_0-\rho-1}{2})P_2(\frac{r-\rho}{2})}.$$

There are 8 possible pairs in this case.

ii) N_0 is type 1) an N_1 is type 11),

$$F(\rho) = \frac{2^{-4+\frac{1}{2}\rho(m-\rho-1)}}{P_2(\frac{\rho-1}{2})P_2(\frac{m_0-\rho-1}{2})P_2(\frac{r-\rho-2}{2})}.$$

There are 32 possible pairs in this case.

$\underline{m = odd,\ r = odd\ (\geq 3)\ and\ \rho = even}$

iii) N_0 is type 2) or 3), and N_1 is type 12) or 13),

$$F(\rho) = \frac{2^{-m+r-3+\frac{1}{2}\rho(m-\rho-3)}(1 + \delta_0 2^{-\frac{\rho}{2}})(1 + \delta_0 2^{-\frac{m_0-\rho}{2}})(1 + \delta_1 2^{-\frac{r-\rho-1}{2}})}{P_2(\frac{\rho}{2})P_2(\frac{m_0-\rho}{2})P_2(\frac{r-\rho-1}{2})}.$$

There are 4 possible pairs in this case.

iv) N_0 is type 4) or 5), and N_1 is type 12) or 13),

$$F(\rho) = \frac{2^{\frac{1}{2}\rho(m-\rho-5)}(1 + \delta_0 2^{-\frac{m_0-\rho-2}{2}})(1 + \delta_1 2^{-\frac{r-\rho-1}{2}})}{P_2(\frac{\rho}{2})P_2(\frac{m_0-\rho-2}{2})P_2(\frac{r-\rho-1}{2})}.$$

There are 8 possible pairs in this case.

v) N_0 is type 6) or 7) , and N_1 is type 14),

$$F(\rho) = \frac{2^{-m+r-4+\frac{1}{2}\rho(m-\rho+1)}(1 + \delta_0{}^{-\frac{\rho-2}{2}})}{P_2(\frac{\rho-2}{2})P_2(\frac{m_0-\rho}{2})P_2(\frac{r-\rho-1}{2})}.$$

There are 32 possible pairs in this case.

vi) N_0 is type 8) and N_1 is type 14),

$$F(\rho) = \frac{2^{-2+\frac{1}{2}\rho(m-\rho-1)}}{P_2(\frac{\rho-2}{2})P_2(\frac{m_0-\rho-2}{2})P_2(\frac{r-\rho-1}{2})}.$$

There are 128 possible pairs in this case.

$m =$ odd, $r =$ even (≥ 2) and $\rho =$ odd.

vii) N_0 is type 9) or 10), and N_1 is type 1),

$$F(\rho) = \frac{2^{-m+r-4+\frac{1}{2}\rho(m-\rho-1)}(1 + \delta_0 2^{-\frac{\rho-1}{2}})}{P_2(\frac{\rho-1}{2})P_2(\frac{m_0-\rho}{2})P_2(\frac{r-\rho-1}{2})}.$$

There are 8 possible pairs in this case.

viii) N_0 is type 11) and N_1 is type 1),

$$F(\rho) = \frac{2^{-4+\frac{1}{2}\rho(m-\rho-1)}}{P_2(\frac{\rho-1}{2})P_2(\frac{m_0-\rho-2}{2})P_2(\frac{r-\rho-1}{2})}.$$

There are 32 possible pairs in this case.

$m =$ odd, $r =$ even (≥ 2) and $\rho =$ even

ix) N_0 is type 12) or 13), and N_1 is type 2) or 3),

$$F(\rho) = \frac{2^{-r-3+\frac{1}{2}\rho(m-\rho-3)}(1 + \delta_0 2^{-\frac{m_0-\rho-1}{2}})(1 + \delta_1 2^{-\frac{\rho}{2}})(1 + \delta_1 2^{-\frac{r-\rho}{2}})}{P_2(\frac{\rho}{2})P_2(\frac{m_0-\rho-1}{2})P_2(\frac{r-\rho}{2})}.$$

There are 4 possible pairs in this case.

x) N_0 is type 12) or 13), and N_1 is type 4) or 5),

$$F(\rho) = \frac{2^{\frac{1}{2}\rho(m-\rho-5)}(1 + \delta_0 2^{-\frac{m_0-\rho-1}{2}})(1 + 2^{-\frac{r-\rho-2}{2}})}{P_2(\frac{\rho}{2})P_2(\frac{m_0-\rho-1}{2})P_2(\frac{r-\rho-2}{2})}.$$

There are 8 possible pairs in this case.

xi) N_0 is type 14) and N_1 is type 6) or 7),

$$F(\rho) = \frac{2^{-r-4+\frac{1}{2}\rho(m-\rho+1)}(1 + \delta_1 2^{-\frac{\rho-2}{2}})}{P_2(\frac{\rho-2}{2})P_2(\frac{m_0-\rho-1}{2})P_2(\frac{r-\rho}{2})}.$$

There are 32 possible pairs in this case.

xii) N_0 is type 14) and N_1 is type 8),

$$F(\rho) = \frac{2^{-2+\frac{1}{2}\rho(m-\rho-1)}}{P_2(\frac{\rho-1}{2})P_2(\frac{m_0-\rho-1}{2})P_2(\frac{r-\rho-2}{2})}.$$

There are 128 possible pairs in this case.

$m =$ even, $r =$ even (≥ 2) and $\rho =$ odd

xiii) N_0 and N_1 are type 1),

$$F(\rho) = \frac{2^{-4+\frac{1}{2}\rho(m-\rho-1)}}{P_2(\frac{\rho-1}{2})P_2(\frac{m_0-\rho-1}{2})P_2(\frac{r-\rho-1}{2})}.$$

There are 16 possible pairs in this case.

$m = $ even, $r = $ even (≥ 2) and $\rho = $ even

xiv) N_0 and N_1 are type 2) or 3),

$$F(\rho) = \frac{2^{-m-2+\frac{1}{2}\rho(m-\rho-1)}(1 + \delta_0 2^{-\frac{\rho}{2}})(1 + \delta_0 2^{-\frac{m_0-\rho}{2}})(1 + \delta_1 2^{-\frac{\rho}{2}})(1 + \delta_1 2^{-\frac{r-\rho}{2}})}{P_2(\frac{\rho}{2})P_2(\frac{m_0-\rho}{2})P_2(\frac{r-\rho}{2})}.$$

There are 4 possible pairs in this case.

xv) N_0 is type 2) or 3), and N_1 is type 4) or 5),

$$F(\rho) = \frac{2^{-(m-r)+1+\frac{1}{2}\rho(m-\rho-3)}(1 + \delta_0 2^{-\frac{\rho}{2}})(1 + \delta_0 2^{-\frac{m_0-\rho}{2}})(1 + \delta_1 2^{-\frac{r-\rho-2}{2}})}{P_2(\frac{\rho}{2})P_2(\frac{m_0-\rho}{2})P_2(\frac{r-\rho-2}{2})}.$$

There are 8 possible pairs in this case.

xvi) N_0 is type 4) or 5), and N_1 is type 2) or 3),

$$F(\rho) = \frac{2^{-r+1+\frac{1}{2}\rho(m-\rho-3)}(1 + \delta_0 2^{-\frac{m_0-\rho-2}{2}})(1 + \delta_1 2^{-\frac{\rho}{2}})(1 + \delta_1 2^{-\frac{r-\rho}{2}})}{P_2(\frac{\rho}{2})P_2(\frac{m_0-\rho}{2})P_2(\frac{r-\rho-2}{2})}.$$

There are 8 possible pairs in this case.

xvii) N_0 and N_1 are type 4) or 5),

$$F(\rho) = \frac{2^{4+\frac{1}{2}\rho(m-\rho-5)}(1 + \delta_0 2^{-\frac{m_0-\rho-2}{2}})(1 + \delta_1 2^{-\frac{r-\rho-2}{2}})}{P_2(\frac{\rho}{2})P_2(\frac{m_0-\rho-2}{2})P_2(\frac{r-\rho-2}{2})}.$$

There are 16 possible pairs in this case.

xviii) N_0 and N_1 are type 6) or 7),

$$F(\rho) = \frac{2^{-m-4+\frac{1}{2}\rho(m-\rho+3)}(1 + \delta_0 2^{-\frac{\rho-2}{2}})(1 + \delta_1 2^{-\frac{\rho-2}{2}})}{P_2(\frac{\rho-2}{2})P_2(\frac{m_0-\rho}{2})P_2(\frac{r-\rho}{2})}.$$

There are 16 possible pairs in this case.

xix) N_0 is type 6) or 7), and N_1 is type 8),

$$F(\rho) = \frac{2^{-(m-r)-2+\frac{1}{2}\rho(m-\rho-1)}(1 + \delta_0 2^{-\frac{\rho-2}{2}})}{P_2(\frac{\rho-2}{2})P_2(\frac{m_0-\rho}{2})P_2(\frac{r-\rho-2}{2})}.$$

There are 48 possible pairs in this case.

xx) N_0 is type 8), and N_1 is type 6) or 7),

$$F(\rho) = \frac{2^{-r-2+\frac{1}{2}\rho(m-\rho-1)}(1 + \delta_1 2^{-\frac{\rho-2}{2}})}{P_2(\frac{\rho-2}{2})P_2(\frac{m_0-\rho-2}{2})P_2(\frac{r-\rho}{2})}.$$

There are 48 possible pairs in this case.

xxi) N_0 and N_1 are type 8),

$$F(\rho) = \frac{2^{\frac{1}{2}\rho(m-\rho-5)}}{P_2(\frac{\rho-2}{2})P_2(\frac{m_0-\rho-2}{2})P_2(\frac{r-\rho-2}{2})}.$$

There are 256 possible pairs in this case.

$m = \ even, \ r = \ odd \ (\geq 3) \ and \ \rho = \ odd$

xxii) N_0 is type 9) or 10) and N_1 is type 9) or 10),

$$F(\rho) = \frac{2^{-m-4+\frac{1}{2}\rho(m-\rho+3)}(1+\delta_0 2^{-\frac{\rho-1}{2}})(1+\delta_1 2^{-\frac{\rho-1}{2}})}{P_2(\frac{\rho-1}{2})P_2(\frac{m_0-\rho}{2})P_2(\frac{r-\rho}{2})}.$$

There are 4 possible pairs in this case.

xxiii) N_0 is type 9) or 10), and N_1 is type 11)

$$F(\rho) = \frac{2^{-(m-r)-4+\frac{1}{2}\rho(m-\rho-1)}(1+\delta_0 2^{-\frac{\rho-1}{2}})}{P_2(\frac{\rho-1}{2})P_2(\frac{m_0-\rho}{2})P_2(\frac{r-\rho-2}{2})}.$$

There are 16 possible pairs in this case.

xxiv) N_0 is type 11) and N_1 is type 9) or 10),

$$F(\rho) = \frac{2^{-r-4+\frac{1}{2}\rho(m-\rho+1)}(1+\delta_1 2^{-\frac{\rho-1}{2}})}{P_2(\frac{\rho-1}{2})P_2(\frac{m_0-\rho-2}{2})P_2(\frac{r-\rho}{2})}.$$

There are 16 possible pairs in this case.

xxv) N_0 and N_1 are type 11)

$$F(\rho) = \frac{2^{-4+\frac{1}{2}\rho(m-\rho-1)}}{P_2(\frac{\rho-1}{2})P_2(\frac{m_0-\rho-2}{2})P_2(\frac{r-\rho-2}{2})}.$$

There are 64 possible pairs in this case.

$m = \ even, \ r = \ odd \ (\geq 3) \ and \ \rho = \ even$

xxvi) N_0 and N_1 are type 12) or 13),

$$F(\rho) = 2^{-4+\frac{1}{2}\rho(m-\rho-5)}(1+\delta_0 2^{-\frac{m_0-\rho-1}{2}})(1+\delta_1 2^{-\frac{r-\rho-1}{2}}),$$

There are 4 possible pairs in this case.

xxvii) N_0 and N_1 are type 14),

$$F(\rho) = \frac{2^{-4+\frac{1}{2}\rho(m-\rho-3)}}{P_2(\frac{\rho-2}{2})P_2(\frac{m_0-\rho-1}{2})P_2(\frac{r-\rho-1}{2})}.$$

There are 64 possible pairs in this case.

Upper bound of $F(\rho)$ for fixed $m(\geq 30)$ and $r(\geq 2)$.

$m =$ odd and $r =$ odd (i) $-$ vi) of the table).

In cases i), ii), iii), v) and vi), $F(\rho)$ increases as ρ increases. In case iv) $F(\rho) \leq F(\rho + 2) \leq \cdots \leq F(r - 3) \leq 2^2 F(r - 1)$. Therefore we have the following upper bound for each case:

i) $2^{-\frac{r}{2}-3+\frac{1}{2}r(m-r)}/P_2(\frac{r-1}{2})P_2(\frac{m_0-r-1}{2})$,

ii) $2^{-m+\frac{3}{2}r-5+\frac{1}{2}r(m-r)}/P_2(\frac{r-3}{2})P_2(\frac{m_0-r+1}{2})$,

iii) $2^{-m+\frac{1}{2}r+1+\frac{1}{2}r(m-r)}/P_2(\frac{r-1}{2})P_2(\frac{m_0-r+1}{2})$,

iv) $2^{-\frac{m}{2}-\frac{3}{2}r+6+\frac{1}{2}r(m-r)}/P_2(\frac{r-1}{2})P_2(\frac{m_0-r+1}{2})$,

v) $2^{-\frac{3}{2}m+\frac{5}{2}r-3+\frac{1}{2}r(m-r)}/P_2(\frac{r-3}{2})P_2(\frac{m_0-r+1}{2})$,

vi) $2^{-\frac{m}{2}+\frac{r}{2}-2+\frac{1}{2}r(m-r)}/P_2(\frac{r-3}{2})P_2(\frac{m_0-r+1}{2})$.

Since $r \leq \frac{1}{2}(m - 1)$ we have

(62) $$F(\rho) \leq 2^{\frac{1}{2}r(m-r-1)-3}/P_2(\frac{r-1}{2})P_2(\frac{m_0-r+1}{2})$$

for any case i) - vi).

$m =$ odd and $r =$ even (vii) $-$ xii) of the table)

In case vii) viii), x), xi) and xii), $F(\rho)$ increases as ρ increases. In case ix), $F(\rho) \leq F(r - 2) \leq 2^3 F(r)$. Therefore we have the following upper bound for each case:

vii) $2^{-\frac{3}{2}m+\frac{3}{2}r-3+\frac{1}{2}r(m-r)}/P_2(\frac{r-2}{2})P_2(\frac{m_0-r+1}{2})$,

viii) $2^{-\frac{m}{2}-4+\frac{1}{2}r(m-r)}/P_2(\frac{r-2}{2})P_2(\frac{m_0-r-1}{2})$,

ix) $2^{-\frac{5}{2}r+3+\frac{1}{2}r(m-r)}/P_2(\frac{r}{2})P_2(\frac{m_0-r-1}{2})$,

x) $2^{-m-\frac{r}{2}+5+\frac{1}{2}r(m-r)}/P_2(\frac{r-2}{2})P_2(\frac{m_0-r+1}{2})$,

xi) $2^{-\frac{r}{2}-3+\frac{1}{2}r(m-r)}/P_2(\frac{r-2}{2})P_2(\frac{m_0-r-1}{2})$,

xii) $2^{-m+\frac{3}{2}r-3+\frac{1}{2}r(m-r)}/P_2(\frac{r-4}{2})P_2(\frac{m_0-r+1}{2})$.

Since $r \leq \frac{1}{2}(m - 1)$, we have the following (63) and (64) in any case vii) - xii):

(63) $$F(\rho) \leq 2^{\frac{1}{2}r(m-r-1)-3}/P_2(\frac{r}{2})P_2(\frac{m_0-r+1}{2}) \text{ for } r \geq 4.$$

and

(64) $$F(\rho) \leq 2^{\frac{1}{2}r(m-r-1)-1}/P_2(\frac{r}{2})P_2(\frac{m_0-r+1}{2})$$
$$= 2^{m-1}/P_2(1)P_2(\frac{m_0-1}{2}) \text{ for } r = 2.$$

m = even and r = even (xiii) − xxi) in the table)

We have the following upper bound of $F(\rho)$ in each case:

xiii) $2^{-\frac{m}{2}+\frac{r}{2}-4+\frac{1}{2}r(m-r)}/P_2(\frac{r-2}{2})P_2(\frac{m_0-r}{2})$,

xiv) $2^{-m-\frac{r}{2}+9+\frac{1}{2}r(m-r)}/P_2(\frac{r}{2})P_2(\frac{m_0-r}{2})$,

xv) $2^{-2m+\frac{3}{2}r+7+\frac{1}{2}r(m-r)}/P_2(\frac{r-2}{2})P_2(\frac{m_0-r+2}{2})$,

xvi) $2^{-\frac{5}{2}r+8+\frac{1}{2}r(m-r)}/P_2(\frac{r}{2})P_2(\frac{m_0-r-2}{2})$ for $r \leq \frac{m-2}{2}$,

   $2^{-\frac{5}{2}r+8+\frac{1}{2}r(m-r)}/P_2(\frac{r-2}{2})P_2(1)$ for $r = \frac{m}{2}$,

xvii) $2^{-m-\frac{r}{2}+10+\frac{1}{2}r(m-r)}/P_2(\frac{r-2}{2})P_2(\frac{m_0-r}{2})$,

xviii) $2^{-m+\frac{3}{2}r-1+\frac{1}{2}r(m-r)}/P_2(\frac{r-2}{2})P_2(\frac{m_0-r}{2})$,

xiv) $2^{-2m+\frac{5}{2}r-2+\frac{1}{2}r(m-r)}/P_2(\frac{r-4}{2})P_2(\frac{m_0-r+2}{2})$,

xx) $2^{-\frac{3}{2}r+\frac{1}{2}r(m-r)}/P_2(\frac{r-2}{2})P_2(\frac{m_0-r-2}{2})$ for $r \leq \frac{m-2}{2}$,

   $2^{-\frac{3}{2}r-2+\frac{1}{2}r(m-r)}/P_2(\frac{r-4}{2})P_2(1)$ for $r = \frac{m}{2}$,

xxi) $2^{-m-\frac{r}{2}+4+\frac{1}{2}r(m-r)}/P_2(\frac{r-4}{2})P_2(\frac{m_0-r}{2})$.

Since $r \leq \frac{m}{2}$, we have the following (65) and (66) in any case xiii) - xxi).

(65)
$$F(\rho) \leq 2^{\frac{1}{2}r(m-r-1)}/P_2(\frac{r}{2})P_2(\frac{m_0-r+2}{2}) \text{ for } r \geq 4$$

and

(66)
$$F(\rho) \leq 2^{4+\frac{1}{2}r(m-r-1)}/P_2(\frac{r}{2})P_2(\frac{m_0-r+2}{2})$$
$$= 2^{m+1}/P_2(1)P_2(\frac{m_0}{2}) \text{ for } r = 2.$$

m = even and r = odd (xxii) − xxvii) in the table)

We have the following upper bound of $F(\rho)$ in each case:

xxii) $2^{-m+\frac{3}{2}r-1+\frac{1}{2}r(m-r)}/P_2(\frac{r-1}{2})P_2(\frac{m_0-r}{2})$,

xxiii) $2^{-2m+\frac{7}{2}r-6+\frac{1}{2}r(m-r)}/P_2(\frac{r-3}{2})P_2(\frac{m_0-r+2}{2})$,

xxiv) $2^{-\frac{r}{2}-3+\frac{1}{2}r(m-r)}/P_2(\frac{r-1}{2})P_2(\frac{m_0-r-2}{2})$ for $r \leq \frac{m-2}{2}$,

   $2^{-\frac{r}{2}-6+\frac{1}{2}r(m-r)}/P_2(\frac{r-3}{2})P_2(1)$ for $r = \frac{m}{2}$,

xxv) $2^{-m+\frac{3}{2}r-5+\frac{1}{2}r(m-r)}/P_2(\frac{r-3}{2})P_2(\frac{m_0-r}{2})$,

xxvi) $2^{-\frac{m}{2}-\frac{3}{2}+2+\frac{1}{2}r(m-r)}/P_2(\frac{r-1}{2})P_2(\frac{m_0-r}{2})$ for $r \leq \frac{m-2}{2}$,

   $2^{-\frac{m}{2}-\frac{3}{2}r+1+\frac{1}{2}r(m-r)}/P_2(\frac{r-1}{2})$ for $r = \frac{m}{2}$,

xxvii) $2^{-\frac{m}{2}-\frac{3}{2}r-2+\frac{1}{2}r(m-r)}/P_2(\frac{r-3}{2})P_2(\frac{m_0-r}{2})$.

Since $r \leq \frac{m}{2}$ we have the following (67) in any case xxii) - xxvii):

(67)
$$F(\rho) \leq 2^{-1+\frac{1}{2}r(m-r-1)}/P_2(\frac{r-1}{2})P_2(\frac{m_0-r+2}{2})$$

Upperbound of $X(N_0, N_1)$ for fixed $m(\geq 30)$ and $r(\geq 2)$

m = odd and r = odd

In this case L is an odd lattice (see §6, type 9) and 10) for $\alpha_2(L)$).

By (41) and (42) we have

$$\prod_{p \neq 2} \alpha_p(L)\alpha_p(N_0)^{-1}\alpha_p(N_1)^{-1}$$
$$= \prod_{p \neq 2} P_p(\frac{m-1}{2})P_p(\frac{m_0-2}{2})^{-1}P_p(\frac{r-1}{2})^{-1}(1-\delta_0 p^{-\frac{m_0}{2}})^{-1}$$
$$< \prod_{p \neq 2}(1-p^{-\frac{m_0}{2}})^{-1}P_p(\frac{r-1}{2})^{-1},$$

where $\delta_0 = (\frac{(-1)^{\frac{m_0}{2}}dN_0}{p})$.

Therefore by (39), (61) and (62) we have

$$X(N_0, N_1) < \prod_p (1-p^{-\frac{m_0}{2}})^{-1}P_p(\frac{r-1}{2})^{-1} \cdot Y(m, r, 2)$$

Since $r \geq 3$ and $m_0 \geq \frac{1}{2}(m+1) \geq 16)$, we have (by (50) and (51))

$$X(N_0, N_1) < 2 \cdot Y(m, r, 2).$$

m = odd and r = even

In this case L is an odd lattice. For $\alpha_2(L)$ see §6, type 9) and 10).

By (41) and (42) we have

$$\prod_{p \neq 2} \alpha_p(L)\alpha_p(N_0)^{-1}\alpha_p(N_1)^{-1}$$
$$= \prod_{p \neq 2} P_p(\frac{m-1}{2})P_p(\frac{m_0-1}{2})^{-1}P_p(\frac{r-2}{2})^{-1}(1-\delta_1 p^{-\frac{r}{2}})^{-1}$$
$$< \prod_{p \neq 2} P_p(\frac{r-2}{2})^{-1}(1-\delta_1 p^{-\frac{r}{2}})^{-1},$$

where $\delta_1 = (\frac{(-1)^{\frac{r}{2}}dN_1}{2})$.

Therefore if $r \geq 4$ we have (by (61) and (63))

$$X(N_0, N_1) = P_2(\frac{r}{2})^{-1}(1 + \delta 2^{-\frac{m-1}{2}})^{-1}\prod_{p \neq 2} P_p(\frac{r-2}{2})^{-1}(1 - \delta_1 p^{-\frac{r}{2}})^{-1}Y(m, r, 2)$$

$$< \frac{3}{4}(1 - 2^{-\frac{m-1}{2}})^{-1}\varsigma(2)\prod_p P_p(\frac{r}{2})^{-1} \cdot Y(m, r, 2)$$

$$< 2.40Y(m, r, 2),$$

(by (50) and (51))

where $\delta = \pm 1$.

If $r = 2$ then we have by (61) and (64)

$$X(N_0, N_1) = P_2(1)^{-1}(1 + \delta 2^{-\frac{m-1}{2}})\prod_{p \neq 2}(1 - (\frac{-dN_1}{p})p^{-1})^{-1} \cdot 2^2 Y(m, 2, 2)$$

$$< 1.49 \cdot 2^2 Y(m, 2, 2).$$

(by (52))

$m = \;even\; and\; r = \;even$

by (41) and (42) we have

$$\prod_{p \neq 2} \alpha_p(L)\alpha_p(N_0)^{-1}\alpha_p(N_1)^{-1}$$

$$= \prod_{p \neq 2} P_p(\frac{m-2}{2})P_p(\frac{m_0-2}{2})^{-1}P_p(\frac{r-2}{2})^{-1}(1 - \delta p^{-\frac{m}{2}})$$

$$\cdot (1 - \delta_0 p^{-\frac{m_0}{2}})^{-1}(1 - \delta_1 p^{-\frac{r}{2}})^{-1}$$

$$< \prod_{p \neq 2} P_p(\frac{r-2}{2})^{-1}(1 - \delta p^{-\frac{m}{2}})(1 - \delta_0 p^{-\frac{m_0}{2}})^{-1}(1 - \delta_1 p^{-\frac{r}{2}})^{-1},$$

where $\delta = (\frac{(-1)^{\frac{m}{2}}}{p})$, $\delta_0 = (\frac{(-1)^{\frac{m_0}{2}}dN_0}{p})$, and $\delta_1 = (\frac{(-1)^{\frac{r}{2}}dN_1}{p})$.

Therefore if L is odd and $r \geq 4$ we have (by (39), (61), (65), §6 type 6) and 7))

$$X(N_0, N_1) < (1 - 2^{-\frac{m-2}{2}})\prod_p P_p(\frac{r}{2})^{-1}\prod_{p \neq 2}(1 - \delta p^{-\frac{m}{2}})$$

$$(1 - \delta_0 p^{-\frac{m_0}{2}})^{-1}(1 - \delta_1 p^{-\frac{r}{2}})^{-1} \cdot 2^3 Y(m, r, 2)$$

$$< 2.43 \cdot 2^3 Y(m, r, 2).$$

(by (50), (51) and (54))

If L is odd and $r = 2$ then we have (by (39), (61), (66) and §6 type 6) and 7))

$$X(N_0, N_1) < (1 - 2^{-\frac{m-2}{2}})^{-1}(1 - 2^{-2})^{-1}$$

$$\prod_{p \neq 2}(1 - \delta p^{-\frac{m}{2}})(1 - \delta_0 p^{-\frac{m_0}{2}})^{-1}(1 - \delta_1 p^{-1})^{-1} \cdot 2^7 Y(m, 2, 2)$$

$$< 1.50 \cdot 2^7 Y(m, 2, 2).$$

(by (50), (52))

If L is even and $r \geq 4$ then we have (by (39), (61), (65) and §6 type 2))

$$X(N_0, N_1) < \prod_p P_p(\frac{r}{2})^{-1} \prod_{p \neq 2}(1 - \delta_0 p^{-\frac{m_0}{2}})^{-1}(1 - \delta_1 p^{-\frac{r}{2}})^{-1}$$

$$\cdot 2^{m+1} Y(m, r, 2)$$

$$< 2.43 \cdot 2^{m+1} Y(m, r, 2).$$

(by (50), (51))

If L is even and $r = 2$ then we have (by (39), (61), (66) and §6 type 2))

$$X(N_0, N_1) < (1 - 2^{-2})^{-1}(1 + 2^{-\frac{m}{2}})^{-1}\varsigma(\frac{m_0}{2}) \prod_{p \neq 2}(1 - \delta_1 p^{-1})^{-1}$$

$$\cdot 2^{m+5} Y(m, 2, 2)$$

$$< 1.50 \cdot 2^{m+5} Y(m, 2, 2).$$

(by (50) and (52))

$\underline{m\ =\ even\ and\ r\ =\ odd}$

By (41) and (42) we have

$$\prod_{p \neq 2} \alpha_p(L) \alpha_p(N_0)^{-1} \alpha_p(N_1)^{-1}$$

$$= \prod_{p \neq 2}(1 - \delta p^{-\frac{m}{2}}) P_p(\frac{m-2}{2}) P_p(\frac{m_0-1}{2})^{-1} P_p(\frac{r-1}{2})^{-1}$$

$$< \prod_{p \neq 2}(1 - \delta p^{-\frac{m}{2}}) P_p(\frac{r-1}{2})^{-1},$$

where $\delta = (\frac{(-1)^{\frac{m}{2}}}{p})$.

Therefore if L is odd we have (by (39), (61), (67) and §6 type 6) and 7))

$$X(N_0, N_1) < (1 - 2^{-\frac{m-2}{2}})^{-1} \prod_{p \neq 2}(1 - \delta p^{-\frac{m}{2}}) \prod_p P_p(\frac{r-1}{2})^{-1}$$

$$2^2 \cdot Y(m, r, 2)$$

$$< 1.95 \cdot 2^2 \cdot Y(m, r, 2).$$

(by (50), (51) and (54))

If L is even then we have (by (39), (61), (67) and §6 Type 2))

$$X(N_0, N_1) < \prod_p P_p(\frac{r-1}{2})^{-1} 2^m Y(m, r, 2)$$

$$< 1.95 \cdot 2^m Y(m, r, 2).$$

(by (51))

The next proposition is a summary of the above computations, along with Proposition 7.1.

Proposition 7.2. i) If L is odd, then

$$X(N_0, N_1) < 2.43 \cdot 2^3 Y(m, r, 2) \quad \text{for } r \geq 3,$$
$$X(N_0, N_1) < 1.50 \cdot 2^7 Y(m, 2, 2) \quad \text{for } r = 2,$$
$$X(N_0, N_1) < 1.01 \cdot 2^{-1} Y(m, 1, 2) \quad \text{for } r = 1.$$

ii) If L is even, then

$$X(N_0, N_1) < 2.43 \cdot 2^{m+1} Y(m, r, 2) \quad \text{for } r \geq 3,$$
$$X(N_0, N_1) < 1.50 \cdot 2^{m+5} Y(m, 2, 2) \quad \text{for } r = 2,$$
$$X(N_0, N_1) < 1.01 \cdot 2^{m-3} Y(m, 1, 2) \quad \text{for } r = 1.$$

Next we examine the function $Y(m, r, 2)$.

Proposition 7.3. Assume $m \geq 38$. Then $Y(m, r, 2) \leq Y(m, 3, 2)$ for any r such that $3 \leq r \leq \frac{m}{2}$.

Proof. Consider the natural logarithm $ln(Y(m, r, 2)/Y(m, r-1, 2))$ for any r with $3 \leq r - 1 < r \leq \frac{m}{2}$. By Stirling's formula (48) we have $ln(Y(m, r, 2))/Y(m, r-1, 2)) < f_m(r)$, where $f_m(r) = \frac{1}{2}(m - 2r)ln2 + \frac{1}{2}(m - 2r + 1)ln\pi - \frac{r}{2} + \frac{1}{2}(r - 1)ln\frac{r}{2} + \frac{1}{2}(m - r + 1) - \frac{1}{2}(m - r)ln\frac{1}{2}(m - r + 1) + \frac{1}{8}$. Since $r \leq \frac{m}{2}$, $f_m''(r) = \frac{1}{2}(\frac{1}{r} - \frac{1}{m-r+1}) + \frac{1}{2}(\frac{1}{r^2} - \frac{1}{(m-r+1)^2}) > 0$ for any such r, computations shows $f_m'(6) > 0$. Therefore $f_m(r)$ is increasing on $[6, \frac{m}{2}]$. On the other hand we have $f_m(\frac{m}{2}) = \frac{1}{2}ln\pi + \frac{1}{2} + \frac{m}{4}(lnm - ln(m + 2)) - \frac{1}{2}ln\frac{m}{4} + \frac{1}{8} < 0$ for any $m \geq 38$.

Therefore we have $f_m(r) < 0$ for any r such that $6 \leq r \leq \frac{m}{2}$ and for any $m \geq 38$.

Next we consider $f_m(5)$ and $f_m(4)$ as functions of m and evaluate them. We have

$$f_m(5) = \frac{1}{2}(m - 10)ln2 + \frac{1}{2}(m - 9)ln\pi + 2ln\frac{5}{2} + \frac{1}{2}(m - 4)$$
$$- \frac{1}{2}(m - 5)ln\frac{1}{2}(m - 4) - 5 + \frac{1}{8},$$
$$\frac{df_m(5)}{dm} = \frac{1}{2}ln2\pi - \frac{1}{2}ln\frac{1}{2}(m - 4) + \frac{1}{2(m - 4)},$$
$$f_m(4) = \frac{1}{2}(m - 8)ln2 + \frac{1}{2}(m - 7)ln\pi + \frac{3}{2}ln2 + \frac{1}{2}(m - 3)$$
$$- \frac{1}{2}(m - 4)ln\frac{1}{2}(m - 3) - 4 + \frac{1}{8},$$
$$\frac{df_m(4)}{dm} = \frac{1}{2}ln2\pi - \frac{1}{2}ln\frac{1}{2}(m - 3) + \frac{1}{2(m - 3)}.$$

Computations show $\frac{df_m(5)}{dm} < 0$, $f_{38}(5) < 0$, $\frac{df_m(4)}{dm} < 0$ and $f_{38}(4) < 0$. Therefore we have $f_m(5) < 0$ and $f_m(4) < 0$ for all $m \geq 38$. This completes the proof. $\square$

Propositon 7.4.

(i) $$Y(m, 3, 2) < Y(m, 1, 2)m^{-3.2} \quad \text{for } m \geq 43$$

(ii) $$Y(m, 2, 2) < Y(m, 1, 2)m^{-1.57} \quad \text{for } m \geq 43.$$

Proof.

(i) $$Y(m, 3, 2)Y(m, 1, 2)^{-1} = 2^{2m-9}\pi^{m-4}/(m - 3)!$$
$$< m^{-3.2} \quad \text{for } m \geq 43.$$

(ii) $$Y(m, 2, 2)Y(m, 1, 2)^{-1} = 2^{\frac{1}{2}(m-4)}\pi^{\frac{1}{2}(m-3)}/\Gamma(\frac{m-1}{2})$$
$$< m^{-1.57} \quad \text{for } m \geq 43.$$

$\square$

We are now ready to prove the following Lemma:

Lemma 7.5. i) Let L be odd unimodular. Then

$$\omega_{R(2)}/\omega(L) < 59.68Y(m, 1, 2) \quad \text{for } m \geq 43.$$

ii) Let L be even unimodular. Then

$$\omega_{R(2)}/\omega(L) < 3.70 \cdot 2^m Y(m, 1, 2) \text{ for } m \geq 144,$$

where $Y(m, 1, 2) = (\sqrt{2\pi})^m/2\Gamma(\frac{m}{2})$.

Proof. i) Since $(N_0)_q$ and $(N_1)_q$ are determined uniquely by their discriminants (92:1 in [8]), the genera G_{N_0} and G_{N_1} are determined by $(N_0)_2$ and $(N_1)_2$. From the table in §6 we can show that if $r = 1$ then there are at most 11 pairs of genera (adding the numbers for $\rho = 0$ and $\rho = 1$ together), if $r = 2$ then there are at most 70 pairs of genera (adding the numbers for $\rho = 0, 1$ and 2 together), and if $r \geq 3$ then there are at most 24^2 pairs of genera for fixed ρ, r and m. Hence we have

$$\omega_{R(2)}/\omega(L) < 11 \cdot 1.01 \cdot 2^{-1}Y(m, 1, 2) + 70 \cdot 1.50 \cdot 2^7 Y(m, 2, 2)$$
$$+ 24^2 \cdot 2.43 \cdot 2^3 Y(m, 3, 2) \prod_{r=1}^{[\frac{m}{2}]}(r + 1)$$
$$< (5.56 + 70 \cdot 1.50 \cdot 2^7 m^{-1.57}$$
$$+ 24^2 \cdot 2.43 \cdot 2^3(\frac{m^2}{8} + \frac{3}{4}m) \, m^{-3.2})Y(m, 1, 2)$$
$$< 59.68Y(m, 1, 2) \quad \text{for } m \geq 43.$$

ii) Similarly we have

$$\omega_{R(2)}/\omega(L) < 11 \cdot 1.01 \cdot 2^{m-3}Y(m, 1, 2)$$
$$+ 70 \cdot 1.50 \cdot 2^{m+5}Y(m, 2, 2) + 24^2 \cdot 2.43 \cdot 2^{m+1}Y(m, 3, 2)(\frac{m^2}{8} + \frac{3m}{4})$$
$$< 2^m(1.39 + 1.38 + 0.93)Y(m, 1, 2) = 3.70 \cdot 2^m Y(m, 1, 2) \text{ for } m \geq 144.\square$$

Remark. If $m \leq 41$ then we can show that our upper bound for $\omega_{R(2)}/\omega(L)$ is greater than 1. If $m = 42$ then $Y(m, 1, 2)m^{-2.97} < Y(m, 3, 2) < Y(m, 1, 2)m^{-2.98}$. With this estimation, an argument similar to that given in Lemma 7.5 does not give good result. Careful computations might give us a good upper bound but we would rather avoid complicated computations and use the restriction that $m \geq 43$.

§8. The estimation of $\omega_{R(q)}/\omega(L)$, $q \neq 2$.

In this section we give an upper bound of $\omega_{R(q)}/\omega(L)$, $q \neq 2$. Let N_0 and N_1 be lattices such that the pair of genera (G_{N_0}, G_{N_1}) is in $G(q, r, \rho)$, where $1 \leq r \leq [\frac{m-1}{q-1}]$ and $0 \leq \rho \leq min\,(r, m_0)$ with $m_0 = m - r(q-1)$. First we evaluate $|I(V_0'(N_0), V_1'(N_1))|\omega(N_0)\omega X(N_0, N_1)$.

Proposition 8.1.

(i) $\displaystyle\prod_{P \nmid q} \beta_P(N_1)^{-1} = 2^{\frac{1}{2}(q-3)}\pi^{\frac{1}{2}(q-1)}q^{-\frac{1}{4}(q+3)}$ for $r = 1$.

(ii) $\displaystyle\prod_{P \nmid q} \beta_P(N_1)^{-1} \leq 2^{\frac{1}{2}(q-3)}\pi^{\frac{1}{2}(q-1)}q^{-\frac{1}{4}(q+3)}\prod_{p \neq q}\prod_{i=2}^{r}(1 - \frac{1}{p^i})^{-\frac{q-1}{2}}$ for $r \geq 2$.

Proof. Let M be a unimodular hermitian S-lattice of rank 1. Then $w(M) = \frac{1}{2q}$. Therefore by Mass formula (2) and Proposition 5.2 we have

$$\frac{1}{2q} = 2 \cdot q^{\frac{1}{2}}(q^{\frac{q-1}{2}} - 1)^{\frac{1}{2}}(2\pi)^{-\frac{q-1}{2}}\prod_{P}\beta_P(M)^{-1}$$

$$= (2\pi)^{-\frac{q-1}{2}}q^{\frac{q-1}{4}}\prod_{\substack{p \neq q \\ P|p \\ P\,remains \\ prime\,in\,E}}(1 + p^{-f_p})^{-t_p}\prod_{\substack{p \neq q \\ P|p \\ P\,splits \\ in\,E}}(1 - p^{-f_p})^{-t_p},$$

where $t_p f_p = \frac{1}{2}(q - 1)$. Hence we have

$$\prod_{P \nmid q}\beta_P(M)^{-1} = \prod_{\substack{p \neq q \\ P|p \\ P\,remains \\ prime\,in\,E}}(1 + p^{-f_p})^{-t_p}\prod_{\substack{p \neq q \\ P|p \\ P\,splits \\ in\,E}}(1 - p^{-f_p})^{-t_p} = 2^{\frac{q-3}{2}}\pi^{\frac{q-1}{2}}q^{-\frac{q+3}{4}}.$$

Since $(N_1)_P$ is unimodular at $P \mid q$ we have $\beta_P(N_1) = \beta_P(M)$ for $r = 1$ at the prime ideal $P \mid q$. This gives the proof of i). As for ii) by Proposition 5.2 we have

$$\prod_{P \nmid q}\beta_P(N_1)^{-1} = \prod_{\substack{p \neq q \\ P|p \\ P\,remains \\ prime\,in\,E}}\prod_{i=1}^{r}(1 - (-1)^i p^{-f_{pi}})^{-t_p} \cdot \prod_{\substack{p \neq q \\ P|p \\ P\,splits \\ in\,E}}\prod_{i=1}^{r}(1 - p^{-f_{pi}})^{-t_p}$$

$$< 2^{\frac{1}{2}(q-3)}\pi^{\frac{1}{2}(q-1)}q^{-\frac{1}{4}(q+3)}\prod_{p \neq q}\prod_{i=2}^{r}(1 - p^{-i})^{-\frac{q-1}{2}}\quad\text{for } r \geq 2.$$

$\square$

If $m_0 \geq 2$ then by mass formulas (1) and (2) we have

$$X(N_0, N_1) = N_{K/\mathbb{Q}}(D(E/K))^{\frac{r(r+1)}{4}} D(K/\mathbb{Q})^{\frac{r^2}{2}}$$

$$\cdot \frac{|I(V_0'(N_0), V_1'(N_1)|(dN_0)^{\frac{m_0+1}{2}} N_{K/\mathbb{Q}}(\delta N_1)^r}{\alpha_q(N_1)\beta_q(N_1)} \cdot \alpha_q(L)$$

$$\pi^{\frac{1}{4}r(q-1)(2m-rq)} \cdot 2^{-\frac{1}{4}r(r+1)(q-1)}$$

$$\prod_{p \neq q} \frac{\alpha_p(L)}{\alpha_p(N_0)\beta_p(N_1)} \cdot \frac{(\prod_{j=1}^{r}(j-1)!)^{\frac{q-1}{2}} \prod_{i=1}^{m_0} \Gamma(\frac{i}{2})}{\prod_{i=1}^{m} \Gamma(\frac{i}{2})},$$

where $\beta_p(N) = \prod_{P|p} \beta_P(N_1)$ for any p.

Let $F_q(\rho) = \frac{|I(V_0'(N_0), V_1'(N_1)|(dN_0)^{\frac{m_0+1}{2}} N_r(\delta N_1)^r}{\alpha_q(N_0)\beta_q(N_1)}$.

Upper bound of $F_q(\rho)$

$m = $ odd and $r = $ odd

Since m and r are odd, m_0 and ρ are also odd. By (43), Proposition 5.9 and (59) we have

$$F_q(\rho) = \frac{1}{2}(1 + \delta_0 q^{-\frac{m_0-\rho}{2}})q^{\frac{1}{2}(m_0\rho+r\rho-\rho^2-r)} P_q(\frac{m_0-\rho}{2})^{-1} P_q(\frac{\rho-1}{2})^{-1} P_q(\frac{r-\rho}{2})^{-1},$$

where $\delta_0 = \pm 1$ (if $\rho = m_0$ then $\delta_0 = 1$).

Let $1 \leq \rho < \rho + 2 \leq min(m_0, r)$. Then we can easily check $F_q(\rho) \leq F_q(\rho + 2)$.

Therefore we have

(68) $\qquad F_q(\rho) \leq \frac{1}{2}(1 + \delta_0' q^{-\frac{m_0-r}{2}})q^{\frac{1}{2}r(m_0-1)} P_q(\frac{m_0-r}{2})^{-1} P_q(\frac{r-1}{2})^{-1}$ for $r \leq m_0$

and

(69) $\qquad F_q(\rho) \leq q^{\frac{1}{2}(m_0-1)} P_q(\frac{m_0-1}{2})^{-1} P_q(\frac{r-m_0}{2})^{-1}$ for $r \geq m_0$,

where $\delta_0' = \pm 1$ (if $r = m_0$ then $\delta_0' = 1$).

$m = $ odd and $r = $ even

Since m is odd and r is even, m_0 is odd and ρ is even. Propositon 5.9, (44) and (60) give

$$F_q(\rho) \leq \frac{1}{2}(1 + \delta q^{-\frac{\rho}{2}})q^{\frac{1}{2}(m_0\rho+r\rho-\rho^2-r)} P_q(\frac{m_0-\rho-1}{2})^{-1} P_q(\frac{r-\rho}{2})^{-1} P_q(\frac{\rho}{2})^{-1},$$

where $\delta = \pm 1$ (if $\rho = 0$ then $\delta = 1$).

It is easy to check $F(\rho) \leq F(\rho + 2)$ for ρ such that $0 \leq \rho < \rho + 2 \leq min(r, m_0)$. Hence we have

(70) $\quad F_q(\rho) \le F_q(m_0 - 1) = \frac{1}{2}(1 + \delta_0 q^{-\frac{m_0-1}{2}})q^{\frac{1}{2}(rm_0+m_0-2r-1)}$

$$\cdot P_q(\frac{r - m_0 + 1}{2})^{-1} P_q(\frac{m_0 - 1}{2})^{-1} \quad \text{for } r > m_0$$

and

(71) $\qquad F_q(\rho) \le F(r) = \frac{1}{2}(1 + \delta_1 q^{-\frac{r}{2}})q^{\frac{1}{2}r(m_0-1)}$

$$\cdot P_q(\frac{m_0 - r - 1}{2})^{-1} P_q(\frac{r}{2})^{-1} \quad \text{for } r < m_0,$$

where $\delta_0 = \pm 1$ and $\delta_1 = \pm 1$ (if $r = 0$ then $\delta_1 = 1$).

$m = even$ and $r = odd$

Since m is even and r is odd, m_0 is even and ρ is odd. We have

$$F_q(\rho) = \frac{1}{2}q^{\frac{1}{2}(m_0\rho+r\rho-\rho^2-r)}P_q(\frac{\rho - 1}{2})^{-1}P_q(\frac{m_0 - \rho - 1}{2})^{-1}P_q(\frac{r - \rho}{2})^{-1}$$

Then we have $F_q(\rho) \le F_q(\rho + 2)$ for any ρ such that $1 \le \rho < \rho + 2 \le min(m_0, r)$. Therefore we have

(72) $\quad F_q(\rho) \le q^{\frac{1}{2}r(m_0-1)}P_q(\frac{r - 1}{2})^{-1}P_q(\frac{m_0 - r - 1}{2})^{-1} \quad \text{for } r < m_0$

and

(73) $\quad F_q(\rho) \le \frac{1}{2}q^{\frac{1}{2}(m_0r+m_0-2-1)}P_q(\frac{m_0 - 2}{2})^{-1}P_q(\frac{r - m_0 + 1}{2})^{-1} \quad \text{for } r > m_0.$

$m = even$ and $r = even$

Since m and r are even, m_0 and ρ are also even. Then we have

$$F_q(\rho) = \frac{1}{2}q^{\frac{1}{2}(m_0\rho+r\rho-\rho^2-r)}(1 + \delta_0 q^{-\frac{m_0-\rho}{2}})(1 + \delta q^{-\frac{\rho}{2}})$$

$$\cdot P_q(\frac{\rho}{2})^{-1}P_q(\frac{m_0 - \rho}{2})^{-1}P_q(\frac{r - \rho}{2})^{-1},$$

where $\delta_0 = \pm 1$ and $\delta = \pm 1$ (if $\rho = m_0$ then $\delta_0 = 1$ and if $\rho = 0$ then $\delta = 1$).

We can easily show that $F(\rho) \le F(\rho + 2)$ for any ρ such that $0 \le \rho < \rho + 2 \le min(r, m_0)$.

Hence we have

(74) $\qquad F_q(\rho) \le (1 + \delta_0' q^{-\frac{m_0}{2}})q^{\frac{1}{2}r(m_0-1)}P_q(\frac{m_0}{2})^{-1}P_q(\frac{r - m_0}{2})^{-1}$

for $r \ge m_0$ and

(75) $\qquad F_q(\rho) \le \frac{1}{2}(1 + \delta_0'' q^{-\frac{m_0-r}{2}})(1 + \delta_1 q^{-\frac{r}{2}})P_q(\frac{r}{2})^{-1}P_q(\frac{m_0 - r}{2})^{-1}$

for $r \leq m_0$, where $\delta_0' = \pm 1, \delta_0'' = \pm 1$ (if $r = m_0$ then $\delta_0'' = 1$) and $\delta_1 = \pm 1$.

Upper bound of $X(N_0, N_1)$ for $m \geq 43$.

Using the functional equation (47) we have

$$X(N_0, N_1) = 2^{1+\frac{1}{4}r(q-1)(2m-rq-3)}\pi^{\frac{1}{4}r(q-1)(2m-rq-1)}$$

$$\cdot q^{\frac{1}{4}r(rq-2r+1)} \cdot F_q(\rho)\alpha_q(L) \cdot \prod_{p \neq q} \frac{\alpha_p(L)}{\alpha_p(N_0)\beta_p(N_1)}$$

$$\cdot (\prod_{j=1}^{r}(j-1)!)^{\frac{q-1}{2}} \cdot (m_0!(m_0+2)!\cdots(m-2)!)^{-1} \quad \text{for } m_0 = m - r(q-1) \geq 2.$$

m = odd and r = odd

If $m_0 = 1$ then $\rho = 1$, $\omega(N_0) = \frac{1}{2}$ and $|I(V_0'(N_0), V_1'(N_1)| = 2$. Therefore $X(N_0, N_1) = \omega(N_1)/\omega(L)$. Since m is odd, L is odd lattice. Therefore by §6 type 9) and 10) (for $\alpha_2(L)$),(41) (for $\alpha_p(L), p \neq 2$), Proposition 5.7 and Proposition 8.1 we have the following.

$$X(N_0, N_1) < 2.01 \cdot 2^{\frac{q-3}{2}}\pi^{\frac{q-1}{2}}q^{-\frac{q+3}{4}}Y(m, 1, q) \text{ for } r = 1 \text{ (therefore } q = m),$$

and

$$X(N_0, N_1) < 2.01 \cdot 2^{-1} \cdot 3^{-\frac{q-1}{2}}e^{\frac{q-1}{4}}\pi^{\frac{3}{2}(q-1)}q^{-\frac{q+3}{4}} \cdot Y(m, r, q) \quad \text{for } r \geq 3$$
$$\text{with } m = r(q-1) + 1.$$

If $m_0 \geq 3$ then using (68) and (69) we get

$$X(N_0, N_1) < 1.51 \cdot 2^{\frac{1}{2}(q-1)}\pi^{\frac{1}{2}(q-1)}q^{-\frac{1}{4}(q+3)}Y(m, 1, q) \quad \text{for } r = 1$$

and

$$X(N_0, N_1) < 1.51 \cdot 3^{-\frac{1}{2}(q-1)}e^{\frac{1}{4}(q-1)}\pi^{\frac{3}{2}(q-1)}q^{-\frac{1}{4}(q+3)}Y(m, r, q) \quad \text{for } r \geq 3.$$

m = odd and r = even

Since m is odd, m_0 is odd. If $m_0 = 1$ then $\rho = 0$, $|I(V_0'(N_0), V_1'(N_1))| = 1$ and $\omega(N_0) = \frac{1}{2}$. Therefore we have $X(N_0, N_1) = \frac{1}{2}\omega(N_1)/\omega(L)$. Using the mass formulas (1) and (2) we have

$$X(N_0, N_1) < 1.01 \cdot 3^{-\frac{1}{2}(q-1)}\pi^{\frac{3}{2}(q-1)}q^{-\frac{1}{4}(q+3)}Y(m, 2, q) \quad \text{for } r = 2$$

(therefore $m = 2q - 1$),

and

$$X(N_0, N_1) < 1.01 \cdot 3^{-\frac{1}{2}(q-1)}e^{\frac{1}{4}(q-1)}\pi^{\frac{3}{2}(q-1)}q^{-\frac{1}{4}(q+3)}Y(m, r, q) \text{ for } r \geq 4.$$

If $2 \leq r < m_0$, then using (71) we have

$$X(N_0, N_1) < 1.51 \cdot 2^{-1} \cdot 3^{-\frac{1}{2}(q-1)}\pi^{\frac{3}{2}(q-1)}q^{-\frac{1}{4}(q+3)}Y(m, 2, q) \quad \text{for } r = 2,$$

and

$$X(N_0, N_1) < 1.26 \cdot 2^{-1} \cdot 3^{-\frac{1}{2}(q-1)} \cdot \pi^{\frac{3}{2}(q-1)}q^{-\frac{1}{4}(q+3)}Y(m, r, q) \quad \text{for } r \geq 4.$$

If $3 \leq m_0 < r$, then using (70) we have

$$X(N_0, N_1) < 1.01 \cdot 3^{-\frac{1}{2}(q-1)}e^{\frac{1}{4}(q-1}\pi^{\frac{3}{2}(q-1)}q^{-\frac{1}{4}(q+3)}Y(m, r, q).$$

$m = even$ and $r = odd$

Since m is even, m_0 is even. If $1 \leq r < m_0$, then using (72) we have the following:
For an odd lattice L,

$$X(N_0, N_1) < 0.37 \cdot 2^{\frac{1}{2}(q-1)}\pi^{\frac{1}{2}(q-1)}q^{1-\frac{1}{4}(q+3)}Y(m, 1, q)$$

$$\text{for } r = 1 \text{ and } m_0 = 2.$$

$$X(N_0, N_1) < 1.51 \cdot 3^{-1} \cdot 2^{-2+\frac{1}{2}(q-1)}\pi^{\frac{1}{2}(q+1)}q^{-\frac{1}{4}(q+3)}$$

$$\text{for } r = 1 \text{ and } m_0 \geq 4.$$

$$X(N_0, N_1) < 1.51 \cdot 3^{-\frac{1}{2}(q+1)}e^{\frac{1}{4}(q-1)}2^{-2}\pi^{2+\frac{3}{2}(q-1)}q^{-\frac{1}{4}(q+3)}Y(m, r, q)$$

$$\text{for } r \geq 3.$$

For an even lattice L,

$$X(N_0, N_1) < 0.54 \cdot 2^{-1+\frac{1}{2}(q-1)}\pi^{\frac{1}{2}(q-1)}q^{1-\frac{1}{4}(q+3)}Y(m, 1, q) \cdot 2^{(q-1)}$$

$$\text{for } r = 1 \text{ and } m_0 = 2.$$

$$X(N_0, N_1) < 1.25 \cdot 3^{-1} \cdot 2^{-2+\frac{1}{2}(q-1)}\pi^{2+\frac{1}{2}(q-1)}q^{-\frac{1}{4}(q+3)}Y(m, 1, q) \cdot 2^{(q-1)}$$

$$\text{for } r = 1 \text{ and } m_0 \geq 4.$$

$$X(N_0, N_1) < 1.25 \cdot 3^{-\frac{1}{2}(q+1)}e^{\frac{1}{4}(q-1)}2^{-2}\pi^{2+\frac{3}{2}(q-1)}q^{-\frac{1}{4}(q+3)}Y(m, r, q) \cdot 2^{r(q-1)}$$

If $2 \leq m_0 < r$ then using (73) we have the following:

For an odd lattice L,

$$X(N_0, N_1) < 0.37 \cdot 3^{-\frac{1}{2}(q-1)}e^{\frac{1}{4}(q-1)}\pi^{\frac{3}{2}(q-1)}q^{1-\frac{1}{4}(q+3)}$$

$$\cdot q^{\frac{1}{2}(m-rq-1)}Y(m, r, q) \text{ for } m_0 = 2,$$

$$X(N_0, N_1) < 1.51 \cdot 3^{-\frac{1}{2}(q+1)}e^{\frac{1}{4}(q-1)}2^{-2}\pi^{2+\frac{3}{2}(q-1)}$$

$$\cdot q^{-\frac{1}{4}(q+3)+\frac{1}{2}(m-rq-1)}Y(m, r, q) \text{ for } m_0 \geq 4.$$

For an even lattice L,

$$X(N_0, N_1) < 0.54 \cdot 3^{-\frac{1}{2}(q-1)}e^{\frac{1}{4}(q-1)}2^{-1}\pi^{\frac{3}{2}(q-1)}$$

$$\cdot q^{1-\frac{1}{4}(q+3)+\frac{1}{2}(m-rq-1)}Y(m, r, q)2^{r(q-1)} \text{ for } m_0 = 2,$$

$$X(N_0, N_1) < 1.25 \cdot 3^{-\frac{1}{2}(q+1)}e^{\frac{1}{4}(q-1)}2^{-2}\pi^{2+\frac{3}{2}(q-1)}$$

$$\cdot q^{-\frac{1}{4}(q+3)+\frac{1}{2}(m-rq-1)}Y(m, r, q)2^{r(q-1)} \text{ for } m_0 \geq 4.$$

$m = even$ and $r = even$

Since m is even, m_0 is also even. If $2 \leq r \leq m_0$, then by (75) we have the following:

For an odd lattice L,

$$X(N_0, N_1) < 2.67 \cdot 3^{-\frac{1}{2}(q-1)}2^{-2}\pi^{1+\frac{3}{2}(q-1)}q^{-\frac{1}{4}(q+3)}Y(m, 2, q)$$
$$\text{for } r = 2, m_0 = 2,$$
$$X(N_0, N_1) < 2.01 \cdot 3^{-\frac{1}{2}(q+1)}2^{-1}\pi^{2+\frac{3}{2}(q-1)}q^{-\frac{1}{4}(q+3)}Y(m, 2, q)$$
$$\text{for } r = 2, m_0 \geq 4,$$
$$X(N_0, N_1) < 1.67 \cdot 3^{-\frac{1}{2}(q+1)}e^{\frac{1}{4}(q-1)}2^{-1}\pi^{2+\frac{3}{2}(q-1)}q^{-\frac{1}{4}(q+3)}Y(m, r, q)$$
$$\text{for } r \geq 4$$

For an even lattice L,

$$X(N_0, N_1) < 3^{-\frac{1}{2}(q-1)}2^{-1}\pi^{1+\frac{3}{2}(q-1)}q^{-\frac{1}{4}(q+3)}Y(m, 2, 2)2^{2(q-1)}$$
$$\text{for } r = 2, m_0 = 2,$$
$$X(N_0, N_1) < 1.67 \cdot 3^{-\frac{1}{2}(q+1)}2^{-1}\pi^{2+\frac{3}{2}(q-1)}q^{-\frac{1}{4}(q+3)}Y(m, 2, m_0)2^{2(q-1)}$$
$$\text{for } r = 2, m_0 \geq 4,$$
$$X(N_0, N_1) < 1.39 \cdot 3^{-\frac{1}{2}(q+1)}e^{\frac{1}{4}(q-1)}2^{-1}\pi^{2+\frac{3}{2}(q-1)}q^{-\frac{1}{4}(q+3)}Y(m, r, q)2^{r(q-1)}$$
$$\text{for } r \geq 4.$$

If $2 \leq m_0 < r$ then by (74) we have the following:

For an odd lattice L,

$$X(N_0, N_1) < 2.67 \cdot 3^{-\frac{1}{2}(q-1)}e^{\frac{1}{4}(q-1)}2^{-2}\pi^{1+\frac{3}{2}(q-1)}$$
$$\cdot q^{-\frac{1}{4}(q+3)}Y(m, r, q) \text{ for } m_0 = 2,$$
$$X(N_0, N_1) < 1.67 \cdot 3^{-\frac{1}{2}(q+1)}e^{\frac{1}{4}(q-1)}2^{-1}\pi^{2+\frac{3}{2}(q-1)}$$
$$\cdot q^{-\frac{1}{4}(q+3)}Y(m, r, q) \text{ for } m_0 \geq 4.$$

For an even lattice L,

$$X(N_0, N_1) < 3^{-\frac{1}{2}(q-1)}e^{\frac{1}{4}(q-1)}2^{-1}\pi^{1+\frac{3}{2}(q-1)}$$
$$\cdot q^{-\frac{1}{4}(q+3)}Y(m, r, q)2^{r(q-1)} \text{ for } m_0 = 2,$$
$$X(N_0, N_1) < 1.39 \cdot 3^{-\frac{1}{2}(q+1)}e^{\frac{1}{4}(q-1)}2^{-1}\pi^{2+\frac{3}{2}(q-1)}$$
$$\cdot q^{-\frac{1}{4}(q+3)}Y(m, r, q) \cdot 2^{r(q-1)} \text{ for } m_0 \geq 4.$$

The next proposition is a summary of the above computations:

Propositon 8.2. i) Let L be odd unimodular. Then

$$X(N_0, N_1) < 2^2 \cdot 3^{-\frac{1}{2}(q-1)}e^{\frac{1}{4}(q-1)}\pi^{\frac{3}{2}(q-1)}q^{-\frac{1}{4}(q+3)}Y(m, r, q)$$
$$\text{for } r \geq 2,$$
$$X(N_0, N_1) < 0.37 \cdot 2^{\frac{1}{2}(q-1)}\pi^{\frac{1}{2}(q-1)}q^{1-\frac{1}{4}(q+3)}Y(m, 1, q)$$
$$\text{for } r = 1 \text{ and } q \geq 5,$$
$$X(N_0, N_1) < 1.51 \cdot 2\pi \cdot 3^{-\frac{3}{2}}Y(m, 1, 3)$$
$$\text{for } r = 1 \text{ and } q = 3.$$

ii) Let L be even unimodular. Then

$$X(N_0, N_1) < 2^2 \cdot 3^{-\frac{1}{2}(q-1)} e^{\frac{1}{4}(q-1)} \pi^{\frac{3}{2}(q-1)} q^{-\frac{1}{4}(q+3)} Y(m, r, q) 2^{r(q-1)}$$

for $r \geq 2$,

$$X(N_0, N_1) < 0.54 \cdot 2^{-1} \cdot 2^{\frac{1}{2}(q-1)} \pi^{\frac{1}{2}(q-1)} q^{1-\frac{1}{4}(q+3)} Y(m, 1, q) 2^{q-1}$$

for $r = 1, q \geq 5$,

$$X(N_0, N_1) < 1.25 \cdot 3^{-1} 2^{-2} \pi^2 2^{\frac{1}{2}(q-1)} \pi^{\frac{1}{2}(q-1)} q^{-\frac{1}{4}(q+3)} Y(m, 1, q) 2^{q-1}$$

for $r = 1, q = 3$.

Proposition 8.3.

$Y(m, r, q) Y(m, 1, 2)^{-1} m^6 < Y(m-1, r, q) Y(m-1, 1, 2)^{-1} (m-1)^6$ for $m - 1 \geq 43$
and $r(q - 1) \leq m - 1$.

Proof. We have

$$\frac{Y(m, r, q) Y(m, 1, 2)^{-1} m^6}{Y(m-1, r, q) Y(m-1, 1, 2)^{-1} (m-1)^6}$$

$$= \frac{(\sqrt{2\pi})^{r(q-1)}}{(m - r(q-1))(m - r(q-1) + 2) \cdots (m-4)(m-2)} \cdot \frac{\Gamma(\frac{m}{2})}{\sqrt{2\pi} \Gamma(\frac{m-1}{2})} \cdot \frac{m^6}{(m-1)^6}$$

$$\leq \frac{(\sqrt{2\pi})^x}{(m - x)(m - x + 2) \cdots (m-2)} \cdot (\frac{m-1}{2})^{\frac{1}{2}} = f(x) \quad \text{(by (47))},$$

where $x = r(q - 1)$. Then $f(x)/f(x - 2) = 2\pi/(m - x)$ shows $f(2) > f(4) > \cdots <$
$f(2[\frac{m-1}{2}])$. On the other hand, we have

$$f(2[\frac{m-1}{2}]) = (\sqrt{2\pi})^{m-1} (\frac{m-1}{2})^{\frac{1}{2}} / 1 \cdot 3 \cdots \cdot (m-2) \text{ for } m = \text{odd}$$

$$f(2[\frac{m-1}{2}]) = (\sqrt{2\pi})^{m-2} (\frac{m-1}{2})^{\frac{1}{2}} / 2 \cdot 3 \cdots \cdot (m-2) \text{ for } m = \text{even}$$

and

$$f(2) = \frac{2\pi}{m-2} \sqrt{\frac{m-1}{2}}.$$

We can easily check $f(2) < 1$ and $f(2[\frac{m-1}{2}]) < 1$. Hence $f(x) < 1$ for all $x \leq$
$m - 1$. □

Proposition 8.4 i) Let $m = r(q - 1)$ and $m \geq 44$. Then we have

$$2^2 \cdot 3^{-\frac{1}{2}(q-1)} e^{\frac{1}{4}(q-1)} \pi^{\frac{3}{2}(q-1)} q^{-\frac{1}{4}(q+3)} Y(m, r, q) \cdot 2^{r(q-1)}$$

$$< m^{-6} Y(m, 1, 2) \text{ for } r \geq 2.$$

ii) Let $m = q - 1$ and $m \geq 44$. Then

$$2^{\frac{1}{2}(q-1)} \pi^{\frac{1}{2}(q-1)} q^{1-\frac{1}{4}(q+3)} Y(m, 1, q) \cdot 2^{(q-1)}$$

$$< m^{-6} Y(m, 1, 2).$$

Proof i) By Stirling's formula (48) we have

$$(76) \quad \sum_{i=1}^{\frac{m-2}{2}} ln\,(2i)! > \frac{1}{2}(m-2)ln\,\sqrt{2\pi} - \frac{1}{2}(m-2)m$$

$$+ (\frac{m-2}{2})^2 ln\,(m-2) - \frac{1}{2}(\frac{m-2}{2})^2 + \frac{1}{2}(\frac{m-2}{2})ln\,(m-2)$$

$$- \frac{1}{4}(m-2) + ln\,2 + 2 > \frac{m^2}{4}(ln\,(m-2) - 1.64) \text{ for } m \geq 44.$$

Therefore $ln2^2 \cdot 3^{-\frac{1}{2}(q-1)}e^{\frac{1}{4}(q-1)}\pi^{\frac{3}{2}(q-1)}q^{-\frac{1}{4}(q+3)}Y(m,r,q)2^{r(q-1)}$ is bounded by the following function $g_m(r)$:

$$g_m(r) = 2\,ln\,2 - \frac{m}{2r}ln\,3 + \frac{m}{4r} + \frac{3m}{2r}ln\,\pi - (\frac{m}{4r}+1)ln\,(\frac{m}{r}+1)$$

$$+ \frac{1}{4}m(m-r+1)ln\,2 + \frac{1}{4}m(m-r-1)ln\,\pi + \frac{1}{4}r(m-r+1)ln(\frac{m}{r}+1)$$

$$+ \frac{1}{2}mln\,\sqrt{2\pi} + \frac{m}{4}rlnr - \frac{3}{8}mr + \frac{1}{4}mln\,r$$

$$+ \frac{1}{16}\frac{mln\,(r-1)}{r} + \frac{7m}{16r} - \frac{m}{2r}ln\,\sqrt{2\pi} - \frac{m^2}{4}(ln\,(m-2) - 1.64).$$

Then

$$\frac{d\,g_m(r)}{dr} = (\frac{m}{4r} - \frac{mln\,(r-1)}{16r^2} - \frac{7m}{16r^2}) + (\frac{1}{r} - \frac{1}{m+r} - \frac{m}{4r(m+r)})$$

$$+ \frac{mln\,3}{2r^2} + \frac{m}{4r^2}ln\,(\frac{m}{r}+1) - \frac{3mln\,\pi}{2r^2} - \frac{mln\,2\pi}{4} - \frac{3}{8}m$$

$$+ \frac{m}{4}ln\,r + \frac{1}{4}(r-1) + \frac{1}{4}(m-2r+1)ln\,(\frac{m}{r}+1)$$

$$> 0 \text{ for } 4 \leq r \leq \frac{m}{13}.$$

If $r > \frac{m}{13}$, then $r = \frac{m}{12}, \frac{m}{10}, \frac{m}{6}, \frac{m}{4}$ or $\frac{m}{2}$.

We can check that $g_m(2), g_m(3), g_m(\frac{m}{13}), g_m(\frac{m}{12}), g_m(\frac{m}{10}), g_m(\frac{m}{6}), g_m(\frac{m}{4})$ are all bounded by $-0.02m^2$ for $m \geq 44$. For example,

$$g_m(2) = 2\,ln\,2 - \frac{m}{4}ln\,3 + \frac{m}{8} + \frac{3}{8}mln\,(\frac{m}{2}+1)$$

$$- \frac{3}{2}ln\,(\frac{m}{2}+1) - \frac{1}{4}mln\,2 + \frac{1}{4}m^2ln\,2\pi - \frac{1}{4}m^2(ln(m-2) - 1.64)$$

$$= m^2(\frac{2\,ln\,2}{m^2} - \frac{ln\,3}{4m} + \frac{1}{8m} + \frac{3}{8m}ln\,(\frac{m}{2}+1) - \frac{3}{2m^2}ln(\frac{m}{2}+1)$$

$$- \frac{1}{4m}ln\,2 + \frac{1}{4}ln\,2\pi - \frac{1}{4}ln\,(m-2) + \frac{1.64}{4})$$

As a function of m,

$$\frac{2\,ln\,2}{m^2} - \frac{ln\,3}{4m} + \frac{1}{8m}$$

$$+ \frac{3}{8m}ln\,(\frac{m}{2}+1) - \frac{3}{2m^2}ln(\frac{m}{2}+1) - \frac{1}{4m}ln\,2 + \frac{1}{4}ln2\pi$$

$$- \frac{1}{4}ln(m-2) + \frac{1.64}{4}$$

is decreasing for $m \geq 44$, and when $m = 44$, the above is less than -0.04. This gives $g_m(2) \leq -0.04m^2$ for $m \geq 44$. Summarizing the above computations, we have $g_m(r) \leq -0.02m^2$ for any $2 \leq r < \frac{m}{2}$.

On the other hand, we can show that

$$-0.02m^2 < ln\,(Y(m, 1, 2)m^{-6}).$$

This gives the proof for i), $r < \frac{m}{2}$.

For $r = \frac{m}{2}$, which gives $q = 3$, the left hand side of the inequality i) is

$$2^2 \cdot 3^{-\frac{5}{2}} e^{\frac{1}{2}} \pi^3 \cdot 2^{\frac{m^2}{8} + \frac{m}{4}} \pi^{\frac{m^2}{8} - \frac{m}{4}} \cdot 3^{\frac{m^2}{16} + \frac{m}{8}} \cdot \prod_{j=1}^{\frac{m}{2}} (j-1)!/2!4! \cdots (m-2)!.$$

We can show directly that this is bounded by $Y(m, 1, 2)m^{-6}$ for $m \geq 44$. (Note that $Y(m, 1, 2) = 2^{\frac{m}{2} - 1} \pi^{\frac{m}{2}}/(\frac{m}{2} - 1)!.$)

(ii) By (76) we have

$$ln\,2^{\frac{1}{2}(q-1)} \pi^{\frac{1}{2}(q-1)} q^{1-\frac{1}{4}(q+3)} Y(m, 1, q) 2^{q-1}$$
$$= \frac{m}{2} ln\,2 + \frac{m}{2} ln\,\pi - \frac{m}{4} ln\,(m+1) + m\,ln\,2$$
$$+ \frac{1}{4} m(m-2) ln\,2 + \frac{1}{4} m^2 ln\,\pi + \frac{m}{4} ln\,(m+1)$$
$$- \sum_{i=1}^{\frac{m-2}{2}} ln\,(2i)!$$
$$< \frac{1}{4} m^2 ln\,2\pi + m\,ln\,2 + \frac{m}{2} ln\,\pi - \frac{1}{4} m^2 (ln\,(m-2) - 1.64).$$
$$< -0.03m^2 < Y(m, 1, 2)m^{-6} \text{ for } m \geq 44.$$

$\square$

Direct computations give the following proposition:

Proposition 8.5.

(i) $$2^2 \cdot 3^{-\frac{1}{2}(q-1)} e^{\frac{1}{4}(q-1)} \pi^{\frac{3}{2}(q-1)} q^{-\frac{q+3}{4}} 2^{r(q-1)} Y(43, r, q)$$
$$< Y(43, 1, 2) 43^{-6} \text{ for } r \geq 2 \text{ with } r(q-1) < 43.$$

(ii) $$2^{\frac{1}{2}(q-1)} \pi^{\frac{1}{2}(q-1)} q^{1-\frac{1}{4}(q+3)} 2^{q-1} Y(43, 1, q)$$
$$< Y(43, 1, 2) 43^{-6} \text{ for } q - 1 < 43.$$

Proposition 8.6. Let $m \geq 43$. Then

(i) $$2^2 \cdot 3^{-\frac{1}{2}(q-1)} e^{\frac{1}{4}(q-1)} \pi^{\frac{3}{2}(q-1)} q^{-\frac{1}{4}(q+3)} Y(m, r, q) 2^{r(q-1)} < Y(m, 1, 2)m^{-6}$$
for any $r \geq 2$ such that $r(q-1) \leq m$

(ii) $$2^{\frac{1}{2}(q-1)} \pi^{\frac{1}{2}(q-1)} q^{1-\frac{1}{4}(q+3)} Y(m, 1, q) \cdot 2^{q-1} < Y(m, 1, 2)m^{-6}$$
for any $q \leq m+1$,

where $Y(m, 1, 2) = (\sqrt{2\pi})^m / 2\Gamma(\frac{m}{2})$.

Proof. Proof is done by induction on m. Propositions 8.3, 8.4 show if Proposition 8.6 is true for $m-1$ then it is also true for m. Proposition 8.5 gives the first step of the induction. □

Lemma 8.7. Let L be a unimodular lattice of rank $m \geq 43$. Then

$$\omega_{R(q)}/\omega(L) < m^{-3}Y(m, 1, 2),$$

where $Y(m, 1, 2) = (\sqrt{2\pi})^m / 2\Gamma(\frac{m}{2})$.

Proof. The class of $(N_0)_p, p \neq q$ and $p \neq 2$, is determined uniquely by the discriminant of N_0 (see 92:1 in [8]). If $P \nmid q$, then $(\mathcal{N}_1)_p$ is unimodular and determined uniquely (see Proposition 3.2 in [14] and Theorem 7.1 in [7]). There are at most 2 possible classes for $(N_0)_q$ (see 92:2 in [8]). For each $(N_0)_q$ the class of $(\mathcal{N}_1)_p$, for $P \mid q$, determined uniquely by the definition (iv) of $G(q, r, \rho)$ and Theorem 8.2 in [7]. There are at most 2 possible classes for $(N_0)_2$ (see 93: 16 in [8]). Therefore, the number of pairs $(G_{N_0}, G_{\mathcal{N}_1})$ in $G(q, r, \rho)$ is at most 4.

By Proposition 8.2 we have

$$X(N_0, \mathcal{N}_1) < 2^2 \cdot 3^{-\frac{1}{2}(q-1)} e^{\frac{1}{4}(q-1)} \pi^{\frac{3}{2}(q-1)} q^{-\frac{1}{4}(q+3)}$$
$$\cdot Y(m, r, q) 2^{r(q-1)} \text{ for } r \geq 2$$

and

$$X(N_0, \mathcal{N}_1) < 2^{\frac{1}{2}(q-1)} \pi^{\frac{1}{2}(q-1)} q^{1-\frac{1}{4}(q+3)}$$
$$\cdot Y(m, 1, q) 2^{q-1} \text{ for } r = 1$$

Therefore by Proposition 8.6, $X(N_0, \mathcal{N}_1) < Y(m, 1, 2)m^{-6}$ for any r, q, ρ and $m \geq 43$ such that $\rho \leq min(r, m_0), m_0 = m - r(q-1)$. Hence we have (by Lemma 3.13)

$$\omega_{R(q)}/\omega(L) < 4 \sum_{r=1}^{[\frac{m-1}{q-1}]} (r+1) \cdot m^{-6} Y(m, 1, 2)$$
$$< m^{-3} Y(m, 1, 2).$$

□

§9. The estimation of $\omega_{IR(q)}/\omega(L)$.

First consider the case $q \neq 4$.

Let $\mathcal{M}$ be a totally positive definite λS-modular hermitian S-lattice of rank $\frac{m}{q-1}$. Mass formulas (1) and (2) give

$$\omega(\mathcal{M})/\omega(L) = 2^{-\frac{1}{4}r(r+1)(q-1)} \pi^{\frac{1}{4}m(m+1)-\frac{1}{4}r(r+1)(q-1)}$$
$$\cdot q^{\frac{1}{4}r(1+rq)} \cdot \prod_p \alpha_p(L) \cdot \prod_p \beta_P(\mathcal{M})^{-1} \cdot \{\prod_{j=1}^r (j-1)!\}^{\frac{q-1}{2}}$$
$$\cdot [\prod_{i=1}^m \Gamma(\frac{i}{2})]^{-1}.$$

By (42) we have $\alpha_p(L) = (1 - p^{-\frac{m}{2}})P_p/\frac{m-2}{2})$ for $p \neq 2$ (note that $2 \mid r$ and $m = r(q-1)$ gives $4 \mid m$). By Proposition 5.8 we have $\beta_q(M) = q^{\frac{1}{2}r(r+1)}P_q(\frac{r}{2})$. By Proposition 8.1, ii) we have $\prod\limits_{p \nmid q} \beta_p(M)^{-1} \leq 2^{\frac{q-3}{2}}\pi^{\frac{1}{2}(q-1)}q^{-\frac{1}{4}(q+3)}\prod\limits_{p \neq q}\prod\limits_{i=2}^{r}(1 - p^{-i})^{-\frac{1}{2}(q-1)}$.

Using these local densities and (53) we obtain

$$\omega(M)/\omega(L) < \alpha_2(L)2^{\frac{1}{4}m(m-r-3)}\pi^{\frac{1}{4}m(m-r-1)}$$
$$\cdot q^{\frac{1}{4}r(rq-2r+1)}2^{-1}3^{-\frac{1}{2}(q-1)}e^{\frac{1}{4}(q-1)}\pi^{\frac{3}{2}(q-1)}$$
$$\cdot q^{-\frac{1}{4}(q+3)}[\prod_{j=1}^{r}(j-1)!]^{\frac{1}{2}(q-1)}(2!4!\cdots(m-2)!)^{-1}$$

If L is odd unimodular, then $\alpha_2(L) = P_2(\frac{m-2}{2})/(1 + \delta 2^{-\frac{m+2}{2}}), \delta = \pm 1$ and if L is even unimodular, then $\alpha_2(L) = 2^m P_2(\frac{m}{2})/(1 + 2^{-\frac{m}{2}})$ (see §6, types 6) , 7) an 2).) Then we obtain (using the defintion (40))

$$\omega(M)/\omega(L) < 1.01 \cdot 2^{-1} \cdot 3^{-\frac{1}{2}(q-1)}e^{\frac{1}{4}(q-1)}\pi^{\frac{3}{2}(q-1)}$$
$$q^{-\frac{1}{4}(q+3)}Y(m,r,q)2^{r(q-1)}.$$

Therefore by Proposition 8.6, i) and Lemma 4.4 we obtain the following Lemma:

Lemma 9.1. Let L be a unimodular lattice. Then $\omega_{IR(q)}/\omega(L) < Y(m,1,2)m^{-6}$, where $Y(m,1,2) = (\sqrt{2\pi})^m/2\Gamma(\frac{m}{2})$ for $q \neq 4$.

Next, consider the case $q = 4$.

(In this case $E = \mathbb{Q}(\varsigma) = \mathbb{Q}(\sqrt{-1})$ and $K = \mathbb{Q}$.)

Proposition 9.2. Let M be a unimodular hermitian S-lattice of rank $\frac{m}{2}$.

$$\prod_{p \neq 2}\beta_p(M)^{-1} < \frac{1}{32}\pi^3 e^{\frac{1}{2}}\prod_{i=3}^{r}(1 - 2^{-i})$$

Proof. By Proposition 5.2 and the fact that p splits in E for $p \equiv 1 \ (mod\ 4)$ and remains prime in E for $p \equiv 3 \ (mod\ 4)$, we have

$$\prod_{p \neq 2}\beta_p(M)^{-1} = \prod_{p \equiv 1(4)}\prod_{i=1}^{r}(1 - p^{-i})^{-1}\prod_{p \equiv 3(4)}\prod_{i=1}^{r}(1 - (-1)^i p^{-i})^{-1}$$
$$< \prod_{p \equiv 1(4)}(1 - p^{-1})^{-1}\prod_{p \equiv 3(4)}(1 + p^{-1})^{-1}\prod_{p \neq 2}\prod_{i=2}^{r}(1 - p^{-i})^{-1}$$
$$< \frac{\pi^3 e^{\frac{1}{2}}}{32}\prod_{i=3}^{r}(1 - 2^{-i})$$

(by (50), (52) and (53))

With the notation of Proposition 9.2, Mass formulas (1) and (2) give

$$\omega(M)/\omega(L) < \frac{\pi^3 e^{\frac{1}{2}}}{32} \prod_{i=3}^{r}(1 - 2^{-i}) \cdot \frac{\alpha_2(L)}{\beta_2(M)} \cdot \prod_{p \neq 2}(1 - (\frac{(-1)^{\frac{m}{2}}}{p})p^{-\frac{m}{2}})P_2(\frac{m-2}{2})$$

$$\cdot \pi^{\frac{1}{4}m(m+1) - \frac{1}{2}r(r+1)} \prod_{j=1}^{r}(j-1)! [\prod_{i=1}^{m} \Gamma(\frac{i}{2})]^{-1}.$$

If L is odd unimodular and $n(M) = S$, then by Propositions 5.13 and 5.14 we have

$$\prod_{i=3}^{r}(1 - 2^{-i})\frac{\alpha_2(L)}{\beta_2(M)} = \frac{P_2(\frac{m-2}{2})}{1 + \delta_2 - \frac{m-2}{2}} \cdot \frac{1}{2P_2([\frac{r-1}{2}])} \prod_{i=3}^{r}(1 - 2^{-i}) < 0.51.$$

If L is even unimodular and $n(M) = 2S$, then by Proposition 5.12 we have

$$\prod_{i=3}^{r}(1 - 2^{-i})\frac{\alpha_2(L)}{\beta_2(M)} = \frac{2^m P_2(\frac{m}{2})}{1 + 2^{-\frac{m}{2}}} \cdot \frac{1}{2^r P_2(\frac{r}{2})} \prod_{i=3}^{r}(1 - 2^{-i}) < 2^{\frac{m}{2}}.$$

Therefore by using (54) we have the following Proposition:

Proposition 9.3. Let M be a unimodular lattice of rank $\frac{m}{2}$. Assume that $nM = S$ if L is odd, and $n(M) = 2S$ if L is even. Then we have the following:

$$\omega(M)/\omega(L) < 1.01 \cdot \frac{\pi^3 e^{\frac{1}{2}}}{32} \cdot 2^{\frac{m^2}{4}} \pi^{\frac{m^2}{8} - \frac{m}{4}} \prod_{j=1}^{\frac{m}{2}}(j-1)!$$

$$\cdot [2!4! \cdots (m-2)!]^{-1}.$$

Lemma 9.4. Let L be a unimodular lattice of rank $m \geq 44$ then

$$\omega_{IR(4)}/\omega(L) < Y(m, 1, 2)m^{-6},$$

where $Y(m, 1, 2) = (\sqrt{2\pi})^m / 2\Gamma(\frac{m}{2})$.

Proof. By Lemmas 4.9 and 4.10 and Proposition 9.3 we have

$$\omega_{IR(4)}/\omega(L) < 2 \cdot 1.01 \cdot \frac{\pi^3 e^{\frac{1}{2}}}{32} \cdot 2^{\frac{m^2}{4}} \pi^{\frac{m^2}{8} - \frac{m}{4}} \prod_{j=1}^{\frac{m}{2}}(j-1)!$$

$$\cdot [2!4! \cdots (m-2)!]^{-1}.$$

Therefore we have

$$\frac{\omega_{IR(4)}}{\omega(L)} \cdot Y(m, 1, 2)^{-1}$$

$$< \frac{1.62 \cdot 2^{\frac{m^2}{4}} \pi^{\frac{m^2}{8} - \frac{m}{4}} \prod_{j=1}^{\frac{m}{2}}(j-1)!}{2!4! \cdots (m-2)!} \cdot \frac{2(\frac{m-2}{2})!}{(\sqrt{2\pi})^m}.$$

Let us denote this upper bound by $g(m)$. Then we obtain

$$\frac{g(m)}{g(m-2)} \cdot \frac{m^6}{(m-2)^6} = \frac{2^{m-3} \pi^{\frac{1}{2}m-2}}{(m-3)!} (\frac{m}{m-2})^6 < 1.$$

Direct computations show that $g(44) \cdot 44^6 < 1$. Therefore we have the Lemma. $\square$

CHAPTER V
PROOF OF THE THEOREMS

§10. Proof of the theorems.

In this section we complete the proof of Theorems 1, 2 and 3.

Since we have

$$\omega'(L) \leq \sum_{\substack{q \, prime \\ 2 \leq q \leq m}} \omega_{R(q)} + \sum_{\substack{q \, odd \, prime \\ (q-1) \mid m \, or \\ q=4}} \omega_{IR(q)},$$

and there are at most $[\frac{m}{2}]$ prime numbers less than $m+1$,

Lemmas 7.5, 8.7, 9.1 and 9.4 give the following:

If L is odd unimodular lattice, then

$$\omega'(L)/\omega(L) < 59.68 Y(m, 1, 2) + \frac{m}{2} \cdot m^{-3} Y(m, 1, 2)$$

$$+ \frac{m}{2} \cdot m^{-6} Y(m, 1, 2) + m^{-6} Y(m, 1, 2)$$

$$< 60 Y(m, 1, 2) = 65 \cdot (\sqrt{2\pi})^m / 2\Gamma(\frac{m}{2})$$

$$< 30(\sqrt{2\pi})^m / \Gamma(\frac{m}{2}) \text{ for } m \geq 43.$$

If L is even unimodular lattice, then

$$\omega'(L)/\omega(L) < (3.70 \cdot 2^m + \frac{m}{2} \cdot m^{-3} + \frac{m}{2} \cdot m^{-6} + m^{-6}) Y(m, 1, 2)$$

$$< 4 \cdot 2^m Y(m, 1, 2) = 4 \cdot 2^m (\sqrt{2\pi})^m / 2\Gamma(\frac{m}{2})$$

$$< 2^{m+1} (\sqrt{2\pi})^m / \Gamma(\frac{m}{2}) \text{ for } m \geq 144.$$

This gives Theorem 2. By estimation of the functions $30(\sqrt{2\pi})^m / \Gamma(\frac{m}{2})$ and $2^{m+1}(\sqrt{2\pi})^m / \Gamma(\frac{m}{2})$ for $m \geq 43$ (resp. $m \geq 144$) we obtain Theorem 1.

Let $\omega''(L) = \sum_{\substack{cls \, K \subseteq G_L}} \frac{1}{|O(K)|}$, where the summation is over the classes of lattices K such that $q \mid |O(K)|$ with some odd prime number q. Then

$$\omega''(L) < \sum_{\substack{q \neq 2 \\ q \, prime \\ q \leq m}} \omega_{R(q)} + \sum_{\substack{q \, odd \, prime \\ (q-1) \mid m}} \omega_{IR(q)}.$$

Therefore Lemmas 8.7 and 9.2 give the proof of Theorem 3.

I would like to express my appreciation to Professor John S. Hsia for suggesting this reserach problem and valuable discussions.

I would like to thank my friend Douglas Brozovic for correcting my English very carefully.

REFERENCES

1. Biermann, J.: Gitter mit kleiner Automorphismengruppe in Geschlechtern von $\mathbb{Z}$ - Gittern mit positive-definiter quadratischer Form (Diss), *Göttingen*, 1981.

2. Böge, S.: Schiefhermitesche Formen über Zahlkörpern und Quaternionen-schiefkörpern, *J. Reine Angew. Math.* 221 (1966), pp. 85-112.

3. Braun, H.: Zur Theorie der hermiteschen Formen, *Abh. Math. Sem. Hamburg* 14 (1941), pp. 61-150.

4. Conway, J. H., et al.: Atlas of Finite Groups, *Clarendon Press, Oxford, 1985,* 1985.

5. Feit, W.: On integral representation of finite groups, *Proc. London Math. Soc.* (3), (1974), pp. 633-683.

6. Hsia, J. S.: Arithmetic theory of integral quadratic forms, *Proceedings of the Queen's Number Theory Conference*, 1979, Ed. P. Ribenboim, vol. 54, pp. 173-204.

7. Jacobowitz, R.: Hermitian forms over local fields, *Am. J. Math.* 84 (1962), pp. 441-465.

8. O'Meara, O. T.: Introduction to Quadratic Forms, *Springer-Verlag*, New York, 1963.

9. O'Meara, O. T.: The construction of indecomposable positive definite quadratic forms, *J. Reine Angew. Math.* 276 (1975), pp. 99-123.

10. Pall, G.: The weight of a genus of positive n -ary quadratic forms, *Proc. Sympos. Pure Math.* VIII (1965) (Amer. Math. Soc., Providence, R.I.), pp. 95-105.

11. Quebbemann, H.G.: Zur Klassifikation unimodularer Gitter mit Isometrie von Primzahlordnung, *J. Reine Angew. Math.* 326 (1981), pp. 158-170.

12. Rehmann, U.: Klassenzahlen einiger totaldefiniter klassischer Gruppen über Zahlkörpern (Diss), *Göttingen*, 1971.

13. Scharlau, W.: Unzerlegbare quadratische Formen, preprint.

14. Shimura, G.: Arithmetic of unitary groups, Annals of Math. 79 (1964), 369-409.

15. Siegel, C. L.: Über die analytische Theorie der quadratischen Formen, Annals of
     Math. 36 (1935), pp. 527-606.

16. Watson, G. L.: The 2-adic density of a quadratic form, Mathematika 23 (1976), pp.
     94-106.

17. Watson, G. L.: Existence of an indecomposable positive quadratic form in a given
     genus of rank at least 14. Acta Arithm. XXXV (1979), pp. 55-100.

Department of Mathematics
The Ohio State University
Columbus, Ohio 43210

MEMOIRS of the American Mathematical Society

SUBMISSION. This journal is designed particularly for long research papers (and groups of cognate papers) in pure and applied mathematics. The papers, in general, are longer than those in the TRANSACTIONS of the American Mathematical Society, with which it shares an editorial committee. Mathematical papers intended for publication in the Memoirs should be addressed to one of the editors:

Ordinary differential equations, partial differential equations and applied mathematics to ROGER D. NUSSBAUM, Department of Mathematics, Rutgers University, New Brunswick, NJ 08903

Harmonic analysis, representation theory and Lie theory to ROBERT J. ZIMMER, Department of Mathematics, University of Chicago, Chicago, IL 60637

Abstract analysis to MASAMICHI TAKESAKI, Department of Mathematics, University of California, Los Angeles, CA 90024

Classical analysis (including complex, real, and harmonic) to EUGENE FABES, Department of Mathematics, University of Minnesota, Minneapolis, MN 55455

Algebra, algebraic geometry and number theory to DAVID J. SALTMAN, Department of Mathematics, University of Texas at Austin, Austin, TX 78713

Geometric topology and general topology to JAMES W. CANNON, Department of Mathematics, Princeton University, Princeton, NJ 08544

Algebraic topology and differential topology to RALPH COHEN, Department of Mathematics, Stanford University, Stanford, CA 94305

Global analysis and differential geometry to JERRY L. KAZDAN, Department of Mathematics, University of Pennsylvania, E1, Philadelphia, PA 19104-6395

Probability and statistics to BURGESS DAVIS, Departments of Mathematics and Statistics, Purdue University, West Lafayette, IN 47907

Combinatorics and number theory to CARL POMERANCE, Department of Mathematics, University of Georgia, Athens, GA 30602

Logic, set theory and general topology to JAMES E. BAUMGARTNER, Department of Mathematics, Dartmouth College, Hanover, NH 03755

Automorphic and modular functions and forms, geometry of numbers, multiplicative theory of numbers, zeta and L-functions of number fields and algebras to AUDREY TERRAS, Department of Mathematics, University of California at San Diego, La Jolla, CA 92093

All other communications to the editors should be addressed to the Managing Editor, RONALD L. GRAHAM, Mathematical Sciences Research Center, AT&T Bell Laboratories, 600 Mountain Avenue, Murray Hill, NJ 07974.

General instructions to authors for

PREPARING REPRODUCTION COPY FOR MEMOIRS

> **For more detailed instructions send for AMS booklet, "A Guide for Authors of Memoirs."**
> **Write to Editorial Offices, American Mathematical Society, P.O. Box 6248,**
> **Providence, R.I. 02940.**

MEMOIRS are printed by photo-offset from camera copy fully prepared by the author. This means that, except for a reduction in size of 20 to 30%, the finished book will look exactly like the copy submitted. Thus the author will want to use a good quality typewriter with a new, medium-inked black ribbon, and submit clean copy on the appropriate model paper.

Model Paper, provided at no cost by the AMS, is paper marked with blue lines that confine the copy to the appropriate size. Author should specify, when ordering, whether typewriter to be used has **PICA**-size (10 characters to the inch) or **ELITE**-size type (12 characters to the inch).

Line Spacing — For best appearance, and economy, a typewriter equipped with a half-space ratchet — 12 notches to the inch — should be used. (This may be purchased and attached at small cost.) Three notches make the desired spacing, which is equivalent to 1-1/2 ordinary single spaces. Where copy has a great many subscripts and superscripts, however, double spacing should be used.

Special Characters may be filled in carefully freehand, using dense black ink, or **INSTANT** ("rub-on") **LETTERING** may be used. AMS has a sheet of several hundred most-used symbols and letters which may be purchased for $5.

Diagrams may be drawn in black ink either directly on the model sheet, or on a separate sheet and pasted with rubber cement into spaces left for them in the text. Ballpoint pen is not acceptable.

Page Headings (Running Heads) should be centered, in CAPITAL LETTERS (preferably), at the top of the page — just above the blue line and touching it.

> LEFT-hand, EVEN-numbered pages should be headed with the AUTHOR'S NAME;

> RIGHT-hand, ODD-numbered pages should be headed with the TITLE of the paper (in shortened form if necessary).

> Exceptions: PAGE 1 and any other page that carries a display title require NO RUNNING HEADS.

Page Numbers should be at the top of the page, on the same line with the running heads.

> LEFT-hand, EVEN numbers — flush with left margin;

> RIGHT-hand, ODD numbers — flush with right margin.

> Exceptions: PAGE 1 and any other page that carries a display title should have page number, centered below the text, on blue line provided.

> > FRONT MATTER PAGES should be numbered with Roman numerals (lower case), positioned below text in same manner as described above.

MEMOIRS FORMAT

> **It is suggested that the material be arranged in pages as indicated below.**
> **Note: <u>Starred items (*)</u> are requirements of publication.**

Front Matter (first pages in book, preceding main body of text).

> Page i — *Title, *Author's name.

> Page iii — Table of contents.

> Page iv — *Abstract (at least 1 sentence and at most 300 words).

> > Key words and phrases, if desired. (A list which covers the content of the paper adequately enough to be useful for an information retrieval system.)

> > *<u>1980 Mathematics Subject Classification</u> (<u>1985 Revision</u>). This classification represents the primary and secondary subjects of the paper, and the scheme can be found in Annual Subject Indexes of MATHEMATICAL REVIEWS beginnning in 1984.

> Page 1 — Preface, introduction, or any other matter not belonging in body of text.

> > Footnotes: *Received by the editor date.
> > Support information — grants, credits, etc.

First Page Following Introduction – Chapter Title (dropped 1 inch from top line, and centered). Beginning of Text.

Last Page (at bottom) – Author's affiliation.